WIRELESS PERSONAL COMMUNICATIONS
The Evolution of
Personal Communications Systems

THE KLUWER INTERNATIONAL SERIES
IN ENGINEERING AND COMPUTER SCIENCE

WIRELESS PERSONAL COMMUNICATIONS
The Evolution of Personal Communications Systems

edited by

Theodore S. Rappaport
Brian D. Woerner
Jeffrey H. Reed
Virginia Polytechnic Institute
& State University

KLUWER ACADEMIC PUBLISHERS
Boston / Dordrecht / London

Distributors for North America:
Kluwer Academic Publishers
101 Philip Drive
Assinippi Park
Norwell, Massachusetts 02061 USA

Distributors for all other countries:
Kluwer Academic Publishers Group
Distribution Centre
Post Office Box 322
3300 AH Dordrecht, THE NETHERLANDS

Library of Congress Cataloging-in-Publication Data

A C.I.P. Catalogue record for this book is available
from the Library of Congress.

TABLE OF CONTENTS

PREFACE

I   WIRELESS NETWORKS AND SERVICES

    1.   DID Trunks Via The Cellular System   1
          Thomas F. Evans

    2.   28 GHz Local Multipoint Distribution Service (LMDS):   7
          Strengths and Challenges
          Scott Y. Seidel and Hamilton W. Arnold

    3.   A Wireless Infrastructure to Support Research Activities   19
          Alex Hills and Richard Hovey

    4.   A CDPD Performance Model for an Intelligent   27
          Transportation Systems Architecture
          Evaluation
          Steven P. Arnold, Russell W. Taylor and Mark A. Wallace

    5.   Capacity of Channel Hopping Channel Stream On   39
          Cellular Digital Packet Data (CDPD)
          Jay M. Jacobsmeyer

    6.   A CSMA/CA Protocol for Wireless Desktop   51
          Communications
          Jim Lansford

II   WIRELESS CDMA

    7.   CDMA Forward Link Capacity in a Flat   59
          Fading Channel
          Louay M.A. Jalloul and Kamyar Rohani

8.   The Effect of Directional Antennas in CDMA   71
     Wireless Local Loop Systems
     Alper T. Erdogan, Ayman F. Naguib, and
     Arogyaswami Paulraj

9.   Decision-Directed Coherent Delay-Locked PN   81
     Tracking Loop for DS-CDMA
     M. Sawahashi and F. Adachi

III ANTENNAS, PROPAGATION AND SYSTEM DESIGN

10. Adaptive Beamforming for Wireless Communications   89
     J. Litva, A. Sandhu, K. Cho, and T. Lo

11. PCS System Design Issues in the Presence of   95
     Microwave OFS
     Thomas T. Tran, Solyman Ashrafi and A. Richard Burke

12. An Evaluation Point Culling Algorithm for   111
     Radio Propagation Simulation Based on the
     Imaging Method
     Satoshi Takahashi, Kazuhito Ishida, Hiroshi Yoshiura
     and Arata Nakagoshi

13. Space vs. Polarization Diversity Gain in 2 GHz PCS 1900 123
     Paul Donaldson, Robert Ferguson, Eric Kmiec and
     Robert Voss

14. Minimization of Outage Probability in Cellular   135
     Communication Systems by Antenna Beam Tilting
     Josef Fuhl and Andreas F. Molisch

15. Performance of RSS-, SIR-based Handoff and   147
     Soft Handoff in Microcellular Environments
     Per-Erik Östling

IV. SIMULATION, MODULATION AND EQUALIZATION FOR
     WIRELESS COMMUNICATIONS

16. On the Use of Signal-To-Noise Ratio Estimation   159
     for Establishing Hidden Markov Models
     William H. Tranter and Otto H. Lee

17. On the Performance of 4-Phase Sequences in   171
     Asynchronous CDMA Systems
     Mark D. Burroughs and Stephen G. Wilson

18. Adaptive MLSE Equalization Forms for Wireless   183
     Communications
     Gregory E. Bottomley and Sandeep Chennakeshu

19. Simulation of RBDS AM Subcarrier Modulation   195
     Techniques to Determine BER and Audio Quality
     *Shaheen Saroor, Robert Kubichek, Jim Schroeder
     and Jim Lansford*

20. Simulation of Reverse Channel of Narrowband PCS   201
     Rade Petrovic, John N. Daigle and Paolo Giacomazzi

INDEX   213

PREFACE

Wireless personal communications, or *wireless* as it is now being called, has arrived. The hype is starting to fade, and the hard work of deploying new systems and services for personal communications is underway. In the United States, the FCC propelled the wireless era from infancy to mainstream with a $7.7 billion auction of 60 MHz of radio spectrum in the 1800/1900 MHz band. With the largest single sale of public property in the history of mankind mostly complete, the resources of the entire world are being called upon to develop inexpensive, rapidly deployable wireless systems and subscriber units for an industry that is adding subscribers at greater than 50% annual rate. This growth is commonplace for wireless service companies throughout the world, and in the U.S., where as many as 7 licensed wireless service providers may be competing for cellular/PCS customers within the next couple of years, differentiators in cost, quality, service, and coverage will become critical to customer acceptance and use. Many of these issues are discussed in the papers included in this book.

Wireless communications is being woven into the fabric of the human condition at a rate which is difficult to fathom. While it has taken standard wireline telephone service nearly 120 years to reach about 700 million inhabitants around the globe, it has taken less than 20 years for wireless to reach the 50 million mark. Whereas the wireline growth rate of 3% is in line with global population growth, the planet's wireless subscription base is doubling every 20 months. In following the heels of the world wide telephone network buildout and the maturation of the personal computer industry, the wireless future has arrived, and one of the greatest challenges now facing service providers and manufacturers is the ability to find and train technical personnel in this rapidly expanding field. It is for this reason that the Mobile & Portable Radio Research Group (MPRG) at Virginia Tech sponsors the annual symposium on wireless personal communications during the first full week of June. Wireless practitioners gather in Blacksburg each year, where they can mingle with students, meet new and old friends, and learn the latest in wireless communications theories and techniques from authors who have submitted peer reviewed papers.

It is a pleasure to present this compilation of papers and lectures presented at the 5th Virginia Tech Symposium on Wireless Personal Communications. Papers that deal with the topics of wireless networks and services, code division multiple access, antennas and propagation, and simulation, modulation and equalization for wireless links are presented here. These works represent new and original research in a number of emerging areas, and provide valuable insight into practical and theoretical issues facing the wireless field.

In **Section I. Wireless Networks and Services**, six papers are presented. Thomas Evans proposes a novel wireless trunked service that is being deployed by Bell South in Latin America. By exploiting capacity on existing cellular carriers, the paper shows that it is possible to instantly deploy an office PBX system instead of waiting weeks or months for conventional wireline service. Scott Seidel and Pete Arnold provide new propagation measurement results and system comparisons for the 28 GHz Local Multipoint Distribution Service (LMDS) that may become an important wideband service by the 21st century. Alex Hills and Richard Hovey demonstrate a new testbed approach by using a wireless infrastructure on the Carnegie Mellon University campus. Cellular Digital Packet Data (CDPD) and Wireless LANs (WLAN) are used as integral parts of the proposed infrastructure. Steve Arnold, Russell Taylor, and Mark Wallace present a new CDPD performance model that can be used to evaluate communication network architectures for emerging intelligent transportation systems. Jay Jacobsmeyer derives a method for computing the capacity of CDPD when used in a hopping and non-hopping mode, and demonstrates the usefulness of selecting CDPD channels based on carrier-to-interference measurements. Jim Lansford takes the reader from the cellular world to the world of the personal deskspace. His paper illustrates a pico-cell wireless protocol that is proposed for communications between peripherals in and around a personal computer or workstation.

Section II, Wireless CDMA contains papers that develop analysis techniques and performance measures for code division multiple access systems. Louay M.A. Jalloul and Kamyar Rohani present analysis and simulation results that show the benefits offered by soft handoff. Using a slow Rayleigh fading channel model, a wide range of system parameters are varied to determine the sensitivity of CDMA capacity. Alper T. Erdogan, Ayman F. Naguib, and Arogyaswami Paulraj of Stanford University present analysis techniques that may be used to determine the signal-to-noise ratio for reverse channel CDMA systems using directional antennas. Capacity gains are readily achievable since the base station antenna patterns may be selected to provide minimum interference for a particular mobile user. M. Sawahashi and F. Adachi present a novel implementation of a direct sequence delay-lock tracking loop based on decision-directed feedback. Tracking loops are required for rapid spreading code acquisition and tracking, and simulations are conducted to demonstrate the bit error rate of the new scheme.

Section III, Antennas, Propagation and System Design includes six papers that cover modern aspects of personal communications system design, and provides results for new antenna and propagation methods and models. J. Litva, A. Sandhu, K. Cho and T. Lo present a wide range of diversity receiver structures that employ adaptive beamforming to improve link performance. Thomas Tran, Solyman Ashrafi and Richard Burke demonstrate diffraction and ray tracing prediction models for urban microcell and PCS systems. Satoshi Takahashi, Kazuhito Ishida, Hiroshi Yoshiura and Arata Nakagoshi present a three-dimensional propagation modeling technique that uses the method of images to reduce computation time. Paul Donaldson, Robert Ferguson, Eric Kmiec and Robert Voss describe field measurements at 1900 MHz to quantify the diversity improvement at the base station, and find that polarization diversity may offer advantages to the conventional space diversity scheme. Josef Fuhl and Andreas F. Molisch use the concept of antenna beam tilting to minimize the outage for users in a cellular system, based on carrier-to-interference supplied by co-channel base stations. Soft handoffs based on radio signal strength and interference measurements are the subject of study in Per-Erik Ostling's paper. In this work, Ostling demonstrates that the use of signal-to-interference ratio measurements by the mobile can reduce the number of required handoffs to half of that required when signal strength alone is used by the base station to determine handoff thresholds.

Section IV of this book treats **Simulation, Modulation, and Equalization for Wireless Communications**. Using a simple two-level finite state Markov approach, William H. Tranter and Otto Lee model the mobile radio channel to estimate the signal to noise ratio during fading states. The resulting model is fast and efficient and lends itself well to simulation. Mark D. Burroughs and Stephen G. Wilson make an important observation that 4-phase sequences in a CDMA system offers better cross correlation properties than standard binary sequences. Their work considers deterministic bit patterns and the resulting correlation properties from psuedorandom codes. Gregory E. Bottomley and Sandeep Chennakeshu present the principles of maximum likelihood sequence estimation (MLSE) and demonstrate an Ungerbock receiver for time-varying multipath channels. Shaheen Saroor, Robert Kubichek, Jim Schroeder and Jim Lansford present a simulation methodology for low data rate radio broadcast data systems (RBDS) that use AM subcarrier signaling. In the final paper of this book, Rade Petrovic, John N. Daigle and Paolo Giacomazzi present results of packet throughput for narrowband PCS systems. By considering multiple users, the authors determine the likelihood of packet collisions, which are shown to be close to field measurements.

With the wireless communications field growing at such a clip, one book cannot hope to cover all of the pertinent topics of research. However, we hope this compilation of papers will serve as a useful source book for many key areas which are of interest to the wireless community. Nothing can replace the benefit of personal interaction with authors or the hallway discussions that occur at each symposium, but this book should provide a glimpse into the types of problems being pursued by researchers and practitioners throughout the world.

This text would not have been possible without the hard work and dedication of many people. The outstanding papers presented here are due to the efforts of our authors, to whom we are most grateful to for not only submitting their work, but for visiting our campus and sharing their results in person. The MPRG Industrial Affiliates board, which is made up of representatives from AT&T, Bellcore, BellSouth, BNR, FBI, Grayson Electronics, GTE, MCI, Motorola, National Semiconductor, Southwestern Bell, Telesis Technologies Laboratory, and Texas Instruments plays a major role in wireless research at our university, and has perpetuated this symposium and collection of papers through their financial and moral support. Jenny Frank, Annie Wade and Kathy Wolfe have once again coordinated the symposium and the preparation of this book in wonderful fashion.

Finally, it is our hope that the wireless community finds this collection of papers useful, and we hope to see you at a future symposium on the campus of Virginia Tech.

WIRELESS PERSONAL COMMUNICATIONS
The Evolution of Personal Communications Systems

DID Trunks Via The Cellular System

Thomas F. Evans

Network Access Strategy
BellSouth Wireless, Inc.
1100 Peachtree Street NE, Room 808
Atlanta, Georgia 30309

Abstract

Wired telephone trunks for Private Branch Exchange (PBX) operation are often in short supply or lacking altogether in developing nation telephone systems. Even large businesses must often settle for individual lines which must be answered by an attendant and transferred to the called party. Congestion on these lines is also a problem in addition to their cumbersome operation. Cellular infrastructure can be installed in a short period of time to provide sufficient capacity and dependable call completion for many individual users, but heretofore the cellular system had nothing to offer business users who needed trunks to connect to their PBX.

BellSouth has conceived a service which permits true trunked operation via cellular and other interconnected wireless systems. The cellular system provides the transport from the Mobile Telephone Switching Office (MTSO) to the business site. This service is connected to the trunk side of the PBX, which reacts exactly as it would if it were connected to a wireline trunk. Operation is virtually transparent to parties on either end of the call, providing both Direct Inward Dialing (DID) and Direct Outward Dialing (DOD). One can appreciate that since no wired connection is required, this service can be installed very rapidly. It also affords the business user immediate portability with no number change if the business remains within the same cellular system.

BellSouth has staged a three month trial of this system in Latin America. Acceptance has been quite favorable. Users have commented positively on both the improved percentage of call completion and call audio quality. Some trial participants had wireline trunks connected to their PBX, and the trial simply added to their capacity. Other participants had no trunks and installed PBXs to trial the new technology. Not only will this technology increase cellular revenues, it will stimulate the PBX market which often languishes because there are few trunks available in specific telephone markets.

1. Introduction:

A Private Branch Exchange (PBX), located at a customer's business, connects to the telephone company's Central Office via trunks. These trunks provide for aggregation of calls such that *any* telephone set that is connected to the PBX can receive or send a call via *any* trunk. Special signaling is sent over the trunk to enable the PBX to separate calls for individual telephone sets.

In many countries outside of the United States, trunked telephone service is time consuming to obtain, often taking months or even years to obtain, depending on the situation of the local telephone company. Even when trunked service is obtained, the conventional wired telephone system often provides inferior service in the form of poor audio quality, long delays in obtaining dial tone and a low percentage of completed calls. Many times large companies must resort to using a large number of individual telephone lines instead of trunks for their communications. The obvious disadvantage of this approach is that one can not have Direct Inward Dialing and operators or attendants must be employed to answer the individual lines and manually transfer all incoming calls to the appropriate parties. With this situation so prevalent, it is of little wonder that cellular telephony has had such a high rate of acceptance. A cellular user can obtain their service literally in one day. The cellular system offers quick, high quality connections, but it still does not offer the full range of services that a PBX installation offers, and it is more costly to operate than a PBX installation, reducing its widespread usage to only those officials of a company who can justify the cost.

Three years ago a businessman and a BellSouth official were walking in downtown Caracas, Venezuela when the businessman pointed to a building and said that he could not rent space in this high rise building because he could not obtain wireline trunked service. Could cellular somehow provide a possible solution? This question provided the impetus for BellSouth engineers to develop a way to use cellular channels as trunks.

BellSouth has developed a patent pending service, called Cellu-Trunk™, that provides true trunked service via the cellular system. With Cellu-Trunk™, a customer can obtain a high quality audio trunk in virtually a day's time anywhere in a cellular service area. Cellu-Trunk™ uses standard cellular transceivers with special interface electronics to connect them to a customer's standard PBX. Neither the cellular transceivers nor the PBX need to be modified in any way. Added equipment at the Mobile Telephone Switching Office (MTSO) permits using the individual cellular channels as trunks. Cellu-Trunk™ provides true trunked service and could even be used to connect two Class 4 toll switches.

2. System Description:

(Bold numbers in parentheses refer to items in Figure 1.)

A telephone customer (1) connected to the Public Switched Telephone Network (PSTN) dials the number, 525-0012, of a telephone connected to a Cellu-Trunk™ equipped PBX (2). The number on the dialed telephone is a number assigned to the cellular carrier even though it is now associated with a wired telephone. The reason for this number assignment is so that the PSTN will properly route this dialed number to the MTSO providing Cellu-Trunk™ service. Once the call arrives at the MTSO, the equipment in the MTSO routes the call to a special piece of equipment (3) which contains the additional intelligence required for Cellu-Trunk™. This equipment is called a Routing Correlator and is basically a small capacity telephone switch. All *wired* PBX station telephones at the customer's location share certain cellular telephone numbers which serve as the "cellular trunks" in the Cellu-Trunk™ system. Typically, a company with 100 extensions will have from five to ten cellular

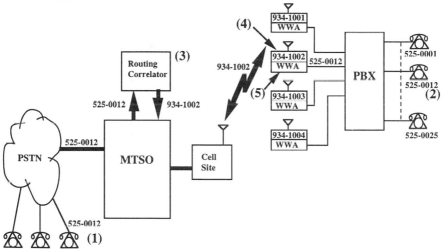

Figure 1 - Block Diagram of Cellu-Trunk

trunks. The Routing Correlator has this information stored in its data storage. When a call arrives at the Routing Correlator, the Routing Correlator looks up to see which cellular trunk can be used to connect this call to its associated PBX. In this case there are four cellular trunk numbers which may be used for this dialed number: 934-1001, 934-1002, 934-1003 and 934-1004. The Routing Correlator also keeps track of which cellular trunks are busy and will not attempt to place a second call to the first cellular trunk, 934-1001, since it is already in use. Once a match is made from the dialed number to an idle cellular trunk, 934-1002, the call for 934-1002 is routed back to the MTSO where it is processed as a standard cellular call.

The MTSO sends out a page for 934-1002. The transceiver (4) at the Cellu-Trunk™ customer's site whose Mobile Identification Number (MIN) is 934-1002 answers the page. This transceiver is connected to a device called a Wired-To-Wireless Adapter, or WWA. The WWA (5) is connected to the transceiver at its handset port and provides all of the functionality of the handset to the transceiver. In this manner a standard, unmodified transceiver may be used in the Cellu-Trunk™ system. The WWA provides the proper response to the cellular system's page and the transceiver answers the call to 934-1002. After the WWA answers the call, the WWA sends a request to the Routing Correlator for the dialed number of the wired telephone. Once the WWA receives the dialed wired telephone number, it seizes the PBX trunk, adds applicable additional signaling (typically E&M 2 or 4 wire signalling) and transmits the dialed number to the PBX. The PBX handles this call in the conventional manner, ringing the appropriate extension. When the extension answers, the PBX provides positive answer supervision to the WWA which, in turn, signals the Routing Correlator that the extension has answered. The Routing Correlator then provides the PSTN with answer supervision, thus completing the telephone call.

As can be expected by the above description, the time required to make an incoming Cellu-Trunk™ call may be slightly longer than the time required to make a wireline trunk call. Many variables enter into the time required, such as MTSO switch design, Cellu-Trunk™ transceiver design, etc. In an effort to mitigate the situation, an announcement feature has been added to the Routing Correlator which allows a business to deliver a personalized message to greet the caller while their call is being processed by Cellu-Trunk™ as soon as a the match between the dialed number and the cellular trunk number is made.

The outgoing Cellu-Trunk™ call is more straightforward. The PBX telephone user dials an access code such as "8" to access the Cellu-Trunk™ system. The user then proceeds to dial the desired number. The PBX collects the dialed digits. After the dialing sequence is completed, the PBX seizes a trunk connected to a WWA. The PBX selects the particular WWA in a prescribed order *opposite* to the order the Routing Correlator assigns WWAs to minimize glare where the MTSO and the PBX vie for the same WWA simultaneously. Once the WWA and the PBX are connected, the PBX sends the dialed number to the WWA. Again the WWA simulates a handset connected to the cellular transceiver and places a standard cellular call to the MTSO.

The MTSO handles this outgoing Cellu-Trunk™ call exactly as it would a standard cellular call, routing it to the PSTN, another Cellu-Trunk™ customer or a cellular customer. The connect time is the same as a standard cellular call. In the instance of an outgoing call, the Routing Correlator is not used; hence, only incoming calls need to be considered when the initial capacity of the Routing Correlator is determined or the Routing Correlator capacity needs to be increased.

There is one factor that must be considered when installing Cellu-Trunk™ which is not significant in normal cellular installations. It may be necessary to use a directional antenna such as a Yagi. The purpose of an antenna of this type is twofold: The directional antenna provides additional gain necessary due to the long antenna runs. The Cellu-Trunk™ equipment will most likely be located in the telephone equipment closet which is often deep within a building. We anticipate antenna cable runs up to 30 meters in length. Also, multiple Cellu-Trunk™ units in an installation will be connected to a four-port antenna coupler, thus reducing the number of antenna cable runs, but introducing another 6 dB of loss. The directional antenna will compensate for these losses. The other reason for the directional antenna is to direct the Cellu-Trunk™ customer toward a specific base station. This consideration is based on system congestion. Since the Cellu-Trunk™ link will see a greater utilization than a typical cellular customer, the cellular operator may want to direct the Cellu-Trunk™ user to a specific cell, particularly if the Cellu-Trunk™ user could communicate with more than one cell as is often the case in larger cities.

3. Advantages:

Cellu-Trunk™ service can be established quickly in markets in which conventional wireline trunks are not available or available only after a long delay. In many instances, the audio quality of the Cellu-Trunk™ system is superior to the PSTN. Another advantage to the Cellu-Trunk™ system that is not readily apparent is that Cellu-Trunk™ affords number portability within a given cellular system. In other words, a company could move within a city served by one cellular system and retain all of their Cellu-Trunk™ numbers. It is not uncommon in some countries to keep a skeleton staff at a company's old location just to answer the telephones and take messages because the company's new location does not yet have telephone service (and will not for come time). With Cellu-Trunk™ **the company would have telephone service as soon as the Cellu-Trunk™ equipment is moved to the new location** and would not need to change their telephone numbers on their business cards, stationery, etc. In fact, a company could have several locations within a city and use the same block of telephone numbers, freely moving individual employees between facilities without having to change their telephone numbers. Individual Cellu-Trunk™s can be added or removed at any time as the user's traffic situation warrants without affecting the remaining Cellu-Trunk™s. All that is required is a translation change at the Routing

Correlator and a minor programming change at the PBX (the PBX programming is identical to the programming required if a wireline trunk is added or deleted).

BellSouth has staged an operational trial of Cellu-Trunk™ in Latin America. Several businesses of different disciplines were selected. Some businesses already had wireline trunk service while others had no existing wireline trunk service. Focus groups and questionnaires before, during and after the trial were utilized to assess customer satisfaction and willingness to pay for this particular service. The response has been favorable from the majority of the trial participants. They report an improved percentage of call completion and better audio. In fact, the improvement in audio is even evident to outside parties calling the Cellu-Trunk™ user. The only negative response is a concern over increased connect time for incoming calls and this concern is being addressed by improved transceiver selection and the personalized announcement feature.

Cellu-Trunk™ provides an important service for businesses who need high quality trunked service in the least amount of time. It adds yet another feature available from the cellular operators in their effort to be the complete communications providers.

28 GHz Local Multipoint Distribution Service (LMDS): Strengths and Challenges

Scott Y. Seidel and Hamilton W. Arnold
Bellcore
331 Newman Springs Road
Room NVC 3X337
Red Bank, NJ 07701
(908) 758-2928
(908) 758-4371 (fax)
seidel@nyquist.bellcore.com

Abstract

Local Multipoint Distribution Service (LMDS) is a terrestrial point-to-multipoint radio service providing wireless broadband access to fixed networks. LMDS could be used to provide wireless access to services ranging from one-way video distribution and telephony to fully-interactive switched broadband multimedia applications. Multiple hub transmitters that communicate via point-to-multipoint radio links with subscribers at fixed locations throughout the service area are placed in a cellular-like layout. LMDS systems attempt to completely reuse the frequency band in each cell through the use of highly directional subscriber antennas and polarization reuse in adjacent cells. Proposed LMDS system architectures differ in terms of cell size, modulation, and hub antenna type. This paper describes the potential service offerings and technical characteristics of LMDS systems. Publicly available system design information was gathered from the FCC Negotiated Rulemaking Committee (NRMC) on co-frequency sharing in the 28 GHz band between LMDS and satellite services. This paper also describes some of the challenges that face LMDS such as a "fragile" propagation environment. The results of a propagation measurement campaign at 28 GHz in Brighton Beach, Brooklyn, to investigate the performance of Local Multipoint Distribution Service (LMDS) indicate that building blockage is a major limitation on the ability to cover a particular location. Strong signals were only received at locations where a line-of-sight path was available between the transmitter and receiver.

1. Potential Services Delivered by LMDS

Recently there has been much interest in providing broadband wireless access to fixed networks via millimeter wave radio transmission in the frequency band between 27.5-29.5 GHz. Local Multipoint Distribution Service (LMDS) could be used to provide wireless access to services ranging from one-way video distribution and telephony to fully-interactive switched broadband multimedia applications. Circuit switched applications such as voice telephony, personal video telephony, backhaul for personal communications services (PCS) and ISDN multimedia services could be accommodated. Packet oriented services such as remote database query, interactive entertainment, personalized information services on virtual channels, transaction processing, and electronic data interchange could also be implemented. Additional potential applications include primary or emergency backup data transport, two-way distance education and corporate training, and high capacity switched data for image transfers and remote consultation for medical users. Interactive uses include video on demand, home shopping, interactive video games, and residential and business data from sub-T1 to multiple-T1 rates. Implementation of digital modulation formats allows the LMDS system provider to take advantage of improvements in digital compression technology and expand to HDTV as these technologies become available.

8

2. Proposed LMDS Systems

2.1 Overview of LMDS

Local Multipoint Distribution Service (LMDS) consists of a cellular-like layout of multiple hub transmitters that communicate via point-to-multipoint radio links with subscribers at fixed locations throughout the service area. System design parameters include antenna patterns, antenna heights, antenna pointing, cell spacing, frequency reuse plan, polarization reuse plan, modulation format, and link budget. A typical arrangement is shown in Figure 1. Cells with omnidirectional hub antennas are nominally circularly shaped and transmit orthogonal linear polarizations (V and H) in adjacent cells. LMDS systems that use sector antennas at the hub employ polarization reuse by alternating orthogonal polarizations in adjacent sectors. The resulting cell shape will be circular when sectors are employed to cover all directions.

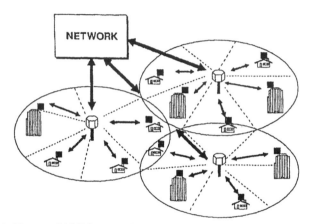

Figure 1. Three cell LMDS system layout with connection to a fixed network (from [2]).

Many LMDS systems attempt to reuse the entire frequency band in each cell. Highly directional subscriber antennas and orthogonal polarization transmissions (polarization reuse) reduce interference levels from adjacent cells. Frequency interleaving could also be used to reduce the amount of inter-cell intra-system interference [1], [3]. With frequency interleaving, the carrier frequencies in different cells are offset by half of the adjacent channel spacing. For FM video with channels spaced 20 MHz apart, the frequency interleave offset is 10 MHz. For FM video channels with an occupancy of 18`MHz, the amount of interference protection obtained from frequency interleaving is estimated to be 10 dB [3]. This technique is most useful with modulation formats such as analog FM which concentrate energy near the center of the channel. Figure 2 shows the polarization/frequency interleaving scheme proposed by CellularVision to achieve reuse of the entire frequency band in each cell. With this arrangement, discrimination in each cell from transmissions from surrounding cells is obtained in polarization, frequency, or both.

For broadcast video distribution, early implementations of LMDS will likely use analog FM with bandwidths on the order of 20 MHz. This bandwidth is wider than the 4.2 MHz required for AM video distribution in order to obtain sufficient FM improvement for high quality video at the cell edge. An analog implementation allows for reduced equipment costs and sufficient channel capacity (50 channels in 1 GHz)

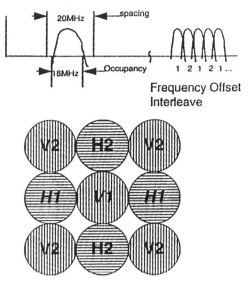

Figure 2. Polarization and frequency interleaving reuse plan for CellularVision LMDS systems (from [2]). Polarization: V or H, Frequency: 1 or 2.

to compete with many current implementations of CATV. Future implementations and evolutions of LMDS will include digital modulation formats to increase channel capacity and provide data and other services mentioned in Section 1.

Link budgets for LMDS systems assume an unblocked path between the hub and subscriber. In many locations, this may not always be the case due to blockage from buildings and trees. Low power active repeaters can be used to fill in areas where there is insufficient signal strength due to excessive blockage along the direct path in the direction of the nearest hub transmitter. A repeater would be located inside the boundaries of a cell at a location where the signal could be received from the hub. The repeater would amplify and redirect the signal in the direction of the coverage hole. The repeater antenna would be cross-polarized to the nearest hub antenna in order to reduce interference to subscribers not utilizing the repeater.

Instead of using a fixed network to interconnect hub locations, point-to-point backbone links could be used. These links could be used to provide source material for video distribution, or interconnection between hubs to a central connection to fixed wireline networks. Detailed system engineering of these links will be required in order to avoid causing interference to other links within the LMDS system.

2.2 Description of LMDS Systems

Table 1 summarizes representative link budgets for LMDS deployments as specified by CellularVision, Video/Phone, and Texas Instruments (TI) [1]. The CellularVision system description is for analog FM modulation, and the Video/Phone and TI descriptions presented here are for digital modulation formats. Both Video/Phone and TI also specified in the NRMC representative systems that can accommodate early entry analog FM modulation. However, those systems are not described here. The CellularVision system description essentially describes the currently operational LMDS system used for broadcast video distribution (wireless cable) in Brighton, Beach, Brooklyn. Table 1 summarizes the link budgets for both hub-to-subscriber (e.g., broadcast video), and future subscriber-to-hub (return path for interactive services) links. Link budgets for the detailed system descriptions are calculated for a subscriber at the edge of the nominal coverage area with an unblocked line-of-sight path to the hub. Both the Video/Phone and TI system descriptions employ power control that can be used to overcome attenuation caused by rain. The amount of rain attenuation that each system is designed to overcome was specified by the system proponent [1]. The actual rain attenuation depends on the rain rate and the length of the radio path through the rain. Note that, in Table 1, the carrier level under rain conditions is calculated under maximum EIRP even if the amount of available power increase over clear sky conditions is greater than the allocated rain fade. The actual carrier level under rain conditions depends upon the algorithm used to implement the power control. The power control tracking loop may either implement an "all or nothing" algorithm, or may more closely track the observed rain attenuation. Both methods would require feedback from subscribers throughout the cell.

2.2.1   CellularVision

The CellularVision system description is for analog FM video distribution with a three mile (4.8 km) cell radius. The link budget allows for a 27.9 dB carrier-to-noise ratio (C/N) at cell edge under clear sky conditions. Under rain attenuation of up to 13 dB, the system is designed for slightly degraded signal quality as long as the video SNR is 42 dB or greater (CCIR "fine" picture quality). The C/N must remain above the threshold for FM improvement due to the capture effect of roughly 8.5 dB. Polarization reuse, directional subscriber antennas, and frequency interleaving are used to reduce intra-system interference. Return links (subscriber-to-hub) require less transmitter power than hub-to-subscriber links due to the much smaller signal bandwidth for low data rate applications such as Pay-Per-View, home shopping, and telephony. These links operate in the guard bands of the downstream broadcast video channels.

2.2.2   Texas Instruments

The Texas Instruments system described in Table 1 uses a four-sector hub antenna to provide overall omnidirectional cell coverage with a nominal radius of 5 km. Power control is implemented to compensate for rain fading. The TI high bit rate digital system with equal data rates in both directions could be used to provide many of the services described in Section 1, including videoconferencing applications. Dedicated spectrum is used for return links. For applications with asymmetrical traffic, users are multiplexed via TDMA.

2.2.3   Video/Phone

The Video/Phone system described here uses a 36-sector hub antenna to provide overall omnidirectional coverage with high reliability over short distances. Extra gain is available in the link budget due to the use of sector antennas at the hub that each have a higher gain than a single omnidirectional antenna. This extra gain and the use of power control allow for increased margin against foliage and rain attenuation. Dedicated spectrum is required for return link transmissions. This system is a high bit rate system that could be used to provide a variety of services described in Section 1.

Table 1. Sample link budget parameters for proposed LMDS systems (from [1]).

Transmitter		Cellular Vision Hub	Video/ Phone 36 Sector Hub	TI Hub	Cellular Vision Sub	Video/ Phone Sub	TI Sub
Modulation	Units	FM Video	QAM 6 Mbps	QPSK 52 Mbps	QPSK 16 kbps	16QAM 45 Mbps	QPSK 52 Mbps
per channel calculations							
Clear Sky Transmitted Power	dBW	-5.0	-15.8	-12.0	-41.0	-20.0	-12.0
Transmitter Antenna Gain	dBi	12.0	29.7	*12.0	31.0	38.0	*34.0
Power Control	dB	0.0	11.0	12.0	0.0	10.0	12.0
EIRP (rain)	dBW	7.0	24.9	12.0	-10.0	28.0	34.0
EIRP (clear sky)	dBW	7.0	13.9	0.0	-10.0	18.0	22.0
Bandwidth	MHz	18	6	52	0.01	22.5	52
Edge of Single Cell Coverage	km	4.8	1.8	5.0	4.8	1.8	5.0
Free Space Path Loss at edge of coverage	dB	135.5	125.9	135.8	135.5	125.9	135.8
Allocated Rain Loss	dB	13.0	5.8	15.0	13.0	5.8	15.0
Receiver Antenna Gain	dBi	31.0	38.0	*34.0	*21.0	29.9	*12.0
Received Power at edge of coverage (rain, full power control)	dBW	-110.5	-68.8	-104.8	-137.5	-73.8	-104.8
Received Power at edge of coverage (clear sky)	dBW	-97.5	-74.0	-101.8	-124.5	-78.0	-101.8
Receiver Noise Figure	dB	6.0	7.5	8.0	3.0	7.5	8.0
Noise Floor	dBW	-125.4	-128.7	-118.8	-161.0	-123.0	-118.8
C/N at edge of coverage (rain, full power control)	dB	14.9	59.9	14.0	23.5	49.2	14.0
C/N at edge of coverage (clear sky)	dB	27.9	54.7	17.0	36.5	45.0	13.0
Required C/(N+I) (rain)	dB	13.0	15.6	13.0	13.0	27.0	13.0
Required C/(N+I) (clear sky)	dB	26.0	15.6	13.0	16.0	27.0	13.0

*NOTE — TI antenna gain and EIRP numbers include specified tolerances of 3 dB at the hub and 1 dB at the subscriber for antenna mispointing and the CellularVision hub employs a separate receive-only sector antenna to increase the margin on subscriber-to-hub return links.

3. Propagation Environment

The major limitation on delivering LMDS to subscribers is the extremely challenging propagation environment at millimeter wave frequencies. As with any radio system, the system performance is limited by the ability of the system to provide sufficient signal strength (minimal path loss) over individual radio links. Attenuation from foliage can be quite large, and wind-induced motion of that foliage can cause significant fluctuations in received signal level. In built-up environments, it is often difficult to provide a line-of-sight path from the hub to each subscriber due to building blockage. Even when line-of-sight paths are available, rain attenuation must be accounted for in link budget design. Atmospheric attenuation is typically not a factor (<0.5 dB) since radio links are established over distances of only a few kilometers. These factors all impact the size of the coverage area that each LMDS hub can serve, and the percentage of customers within that area that can receive sufficient signal strength for high quality service. To investigate the severity of building blockage, and the possibility that blocked locations could be served by signals reflected from buildings, a propagation measurement campaign was conducted at 28 GHz in Brighton Beach, Brooklyn. Buildings within 5 km of the transmitter were mostly 2-3 story residences with clusters of apartment buildings that were 6 or more stories tall. Terrain variations were minimal.

3.1 Propagation Measurements

The vertically polarized broadcast CellularVision signal was used as the transmitted signal source for the propagation measurements. A custom measurement receiver was constructed to block downconvert the entire 1 GHz RF bandwidth and measure the signal strength of a single channel. A 23 dBi standard gain horn was used for the receiver antenna to provide a reasonable compromise between gain for increase of receiver sensitivity for signal detection, and beamwidth for reduced sensitivity to pointing errors. Signal strength was measured at antenna heights over a range of 3.4 meters to 11.3 meters above ground level to determine subscriber antenna heights required to provide service. While the presence of buildings in an urban multipath environment creates coverage holes due to blockage, those same buildings could also reflect signals to other locations within and outside of the cell. Specular reflections were investigated by sweeping the directional receive antenna through 360 degrees azimuth at a fixed elevation angle parallel to the ground. As the antenna was rotated, signal peaks occurred in the directions of multipath arrivals. The minimum detectable signal of these measurements is comparable to the signal level required for "acceptable" video quality in the CellularVision LMDS system when measurements are made under clear sky conditions. Measurements were recorded at seventy-seven discrete locations within 6 km of the CellularVision transmitter, which was located on top of a 95 meter tall apartment building in Brighton Beach, Brooklyn. Measurement locations were selected to cover a wide range of distances and angles from the transmitter that simulate subscriber locations.

3.2 Propagation Measurement Results

For each measurement, the maximum received signal level was computed regardless of whether the signal was the direct or a reflected signal. Figure 3, shows a scatter plot of maximum co-polarized received signal level at the input of the receiver antenna vs. the log of the distance from the transmitter for a receiver antenna height of 13.3 m. This power level refers to the power that would be received by an isotropic antenna. The power received at the output of a directional antenna would be obtained by adding the boresight gain of the antenna to the received isotropic power at the input of the antenna. The horizontal straight line represents the minimum discernible signal level of -111 dBm, and the diagonal straight line corresponds to free space propagation for the CellularVision hub transmitter parameters in Table 1.

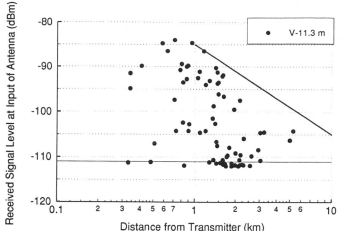

Figure 3. Scatter plot of received co-polarized signal strength vs. log distance from transmitter for a receiver antenna 13.3 meters above ground level.

LMDS coverage areas are generally computed assuming a line-of-sight path between the hub and each subscriber within the cell. However, as indicated in Figure 3, most locations in this built-up environment have signal strengths much weaker than that predicted by free space propagation, even with the receiver antenna 13.3 meters above ground level. The received signal at a majority of locations suffers additional path loss due to blockage by buildings or foliage. In fact, there is a cluster of measured data over the full range of transmitter-receiver separation distances between 300 meters and 5 kilometers with maximum received signal level around -112 dBm at the input to the receiver antenna. Each of these measurements does not represent signal, but indicates that only noise was received at that location. Hence, a coverage hole exists at that location for the antenna height indicated by the symbol. Even though sufficient, albeit not line-of-sight, signal strength is present at some locations over 5 km from the transmitter, there are a large number of locations within the cell where sufficient signal strength is not present, even within 1 km from the transmitter. The presence or absence of a line-of-sight path between the transmitter and receiver significantly impacted the maximum received signal level at any given receiver location and antenna height.

Table 2 summarizes the estimated probability of blockage as a function of antenna height and distance from the transmitter. The values in Table 2 were computed as the percentage of measurement locations where sufficient signal strength was not present (from any direction). As the antenna height was reduced from 13.3 meters to 3.4 or 4.0 meters, the probability of blockage increased from 32% to 74% over all measurement locations. Even within one kilometer of the transmitter, 14% of the locations were blocked at an antenna height of 13.3 meters, and 52% of locations were blocked for antennas 4.0 meters above ground level.

14

Table 2. Percentage of locations where sufficient signal strength was NOT received for different antenna heights and ranges of distances from the transmitter.

Antenna Height	All Measurement Locations	< 3 km From Transmitter	<2 km From Transmitter	<1 km From Transmitter
13.3 meters	32%	32%	28%	14%
7.3 meters	54%	55%	50%	29%
3.4, 4.0 meters	74%	73%	70%	52%

At many of the measurement locations, the received signal level decreased when the antenna height was reduced from 11.3 to 7.3 to 3.4 (4.0) meters. At each measurement location, the receiver antenna was rotated 360 degrees in azimuth to search for signals that arrived via specular reflection at azimuth angles other than that of the direct path from the transmitter. Figure 4 shows the measured signal strength as a function of azimuth angle for three different antenna heights when the measurement van was on Emmons Avenue between 26th Street and 27th Street. The measured data are presented in terms of received power at the input to the antenna. Each curve represents a different antenna height as indicated in the legend. The distance from the transmitter is given just above the legend, and the arrow in the center of the figure indicates the direction of the transmitter from the receiver. In Figure 4, the maximum signal is in the direction of the transmitter for an antenna height of 11.3 meters, and signals arriving from other directions were also detected at all three antenna heights. At many receiver azimuth angles, the signal level received via reflection is greater at the lower antenna heights than when the mast was fully extended. This measurement location was in a business district close to the transmitter with two to three story buildings along Emmons Avenue in the direction away from the transmitter. At 11.3 meters above ground, the antenna was higher than the buildings in the immediate vicinity of the receiver and no reflections were observed from those directions. As the antenna was lowered "into the clutter", reflections from the buildings across the street were detected and the direct signal disappeared. The large specular reflection that was detected only at the high antenna height was

Emmons Av. between 26th St. and 27th St.
Rcvd. Co-Polarized Signal Level at Antenna Aperture (dBm)

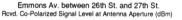

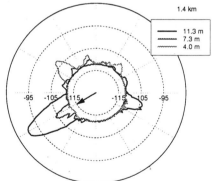

Figure 4. Measured signal strength as a function of azimuth angle for three different antenna heights. Co-polarized specular reflections are distinctly visible at this measurement location.

caused by a distant scatterer that became obstructed from the receiver as the antenna was lowered. This data clearly shows the dependence of signal level on azimuth angle and receiver antenna height.

Increased blockage due to a reduction in receiver antenna height was also seen at many other measurement locations. However, not all measurement locations exhibit the same severe blockage effects due to antenna height as shown in Figure 4. At a few locations, in fact, the maximum received signal level was up to 6 dB stronger at one of the lower antenna heights. In locations with line-of-sight to the transmitter, the received signal level in the direction of the transmitter is unaffected by receiver antenna height. Hence, the impact of subscriber antenna height on received signal level may often be estimated by visual inspection of the specific receiver antenna location to determine whether buildings and/or foliage may obstruct the direct path at some antenna heights and not at others.

3.3   Summary of Propagation Measurements

A propagation measurement campaign was conducted at 28 GHz to investigate the characteristics of millimeter-wave radio channels in a residential urban environment for Local Multipoint Distribution Service (LMDS). These measurements were conducted in a single built-up LMDS environment where multi-story apartment buildings and houses up to three stories tall existed within the coverage area of the transmitter. The measurements here indicate that building blockage is a major limitation on the ability to provide service to a particular location. Strong signals were primarily received only at locations where a line-of-sight path was available between the transmitter and receiver. At many locations, the signal disappeared completely as the antenna was lowered below the roofline of a building directly between the transmitter and receiver. In different environments, the propagation characteristics may be quite different, but the ability to provide sufficient signal strength to subscribers in a particular area is likely strongly related to the ability to provide a line-of-sight path to locations throughout that environment.

4.   Intrasystem Interference

LMDS systems attempt to reuse the entire frequency band in each cell. While polarization reuse and highly directional subscriber antennas are specified, it is unclear how effective these techniques are for different deployment scenarios. Thus, co-frequency transmissions from an adjacent cell may introduce interference into LMDS receivers and degrade system performance, even though the carrier-to-noise ratio is above threshold. Frequency interleaving in adjacent cells provides added protection against self-interference for modulation types such as analog FM video where the signal energy is peaked in the center of the occupied bandwidth. Frequency interleaving is likely less effective for digital modulation techniques that spread the energy across the entire bandwidth. Repeaters are specified for areas where subscribers are blocked from hub transmissions, and backbone links may also be used to interconnect hubs. The impact of these additional transmissions within an LMDS service area on the amount of intrasystem interference is yet to be quantified.

5.   FCC Regulation and Interference from Fixed Satellite Service

Apart from the physical limitations on providing LMDS, regulatory limitations are also an important factor in the growth of LMDS. Use of the 27.5-29.5 GHz frequency band has been requested by mobile satellite service (MSS) providers (for feeder links to satellites providing mobile service) and fixed satellite service (FSS) providers (fixed location subscriber earth station). Point-to-point microwave equipment manufacturers have also requested allocation of the band for point-to-point microwave service. As a result of the complex interference scenarios associated with co-frequency sharing of the band, the FCC formed a Negotiated Rulemaking Committee (NRMC) to study the issues and make recommendations to the Commission for sharing the frequency band between LMDS and satellite services. The Committee investigated four primary interference scenarios covering interference from FSS earth station and MSS feeder link uplinks into LMDS receivers and interference from LMDS transmissions into FSS and MSS satellite receivers. Satellites in both geo-synchronous (GSO) and non-geo-synchronous (NGSO) orbits were considered. The scenario which showed the greatest potential for harmful inter-system interference was for FSS earth stations interfering with LMDS receivers.

Proposed FSS systems plan to use the 28 GHz band for earth station uplinks to satellite receivers. Both LMDS and FSS system proponents envision widespread distribution of earth terminals throughout the same geographic areas. Hence, it is likely that these terminals will be located close to one another, potentially causing harmful interference. Although satellites may be in non-geosynchronous orbit, the fixed portion of FSS refers to the subscriber earth terminals which may be transportable, but remain fixed while in communication with the satellite. Figure 5 shows the desired LMDS signal paths and the FSS interference paths for an FSS earth station accessing a satellite. The interference power generated is directly proportional to the FSS earth station antenna sidelobe level in the direction of LMDS receivers. A single FSS earth station transmitter can simultaneously interfere with many different LMDS receivers. This problem is magnified when the FSS uplink undergoes rain attenuation since proposed system designs implement power control to adaptively increase the transmitted power under heavy rain conditions. At the time this paper is being prepared, the FCC is struggling with how to allocate the 28 GHz band. Options include full allocation of the band for either satellite or terrestrial services, or band segmentation where both satellite and LDMS services would receive less bandwidth than desired.

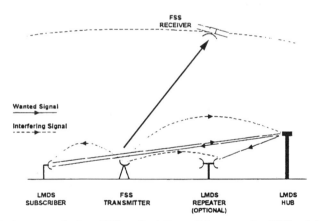

Figure 5. Interference paths from FSS earth station uplinks accessing GSO satellite receivers into LMDS receivers (from [1]).

6. Summary

Local Multipoint Distribution Service (LMDS) can potentially be used to deliver to the end user a variety of services from one-way broadcast video to fully-interactive multimedia applications. Multiple LMDS hubs are arranged in a cellular-like fashion to provide service to subscribers located throughout the service area. Polarization reuse, highly directional subscriber antennas, and frequency interleaving can be used to attempt complete frequency reuse in each cell. Several LMDS system designs are described in this paper. However, the paper describes measurements suggesting that the propagation environment tends to limit the size of coverage areas due to building and foliage blockage and rain attenuation. Repeaters can be used to "fill-in" coverage in blocked locations, but transmissions from repeaters and hub-to-hub backbone links may generate unacceptable levels of intrasystem interference. In addition, the regulatory structure for the 28 GHz band must be determined before LMDS deployment can begin. If these limitations can be overcome, LMDS could play a major role in the deployment of future telecommunications networks.

7. References

[1] Report of the LMDS/FSS 28 GHz Band Negotiated Rulemaking Committee, September 23, 1994.

[2] Documents submitted to the FCC Negotiated Rulemaking Committee, July 27-September 23, 1994.

[3] E.N. Barnhart, R.L. Freeman, B.B. Bossard, "Frequency Reuse in the Cellular LMDS," FCC submission for inclusion in LMDS Rulemaking Record, January 6, 1994.

8. Acknowledgments

The efforts of Roy Murray, Larrie Sutliff, Nelson Sollenberger, Carl Lundgren, Daniel Devasirvatham, Steven Ungar, Anita Krishnan, and Art Daughtry contributed to the success of the measurement campaign and the collection of LMDS system information, and are greatly appreciated.

A Wireless Infrastructure to Support Research Activities

Alex Hills
Distinguished Service Professor
Department of Engineering and Public Policy
Carnegie Mellon University
Pittsburgh, PA 15213
alex.hills@andrew.cmu.edu

Richard Hovey
Member of Technical Staff
Bell Communications Research
Morristown, NJ 07962
hovey@bellcore.com

I. Abstract

In order to support the development of software which will allow seamless access to multiple wireless data networks, we are building a wireless data infrastructure. It will allow Carnegie Mellon researchers and other members of the campus community to use mobile computers to gain access to data networks while they are on-campus or when they are off-campus in the greater Pittsburgh area. The wireless infrastructure, described in this paper, will initially include two wireless networks, each of which will provide access to our campus computer network.

On campus we are building a high speed wireless infrastructure by using wireless local area network (wireless LAN) technology to provide high speed (2 Mbps) service over a portion of the campus. Outside the wireless LAN service area, we are using the cellular digital packet data (CDPD) service offered by Bell Atlantic Mobile. With both the CDPD and wireless LAN networks operational, properly equipped mobile computers will be able to use both networks.

II. Wireless Research at CMU

A number of wireless research projects are underway at Carnegie Mellon's Information Networking Institute. The work within our Wireless Initiative is based on the notion that it is unlikely that a single wireless network able to meet all mobile computing needs will evolve. It is far more probable that many wireless networks (including wireless local area networks) will be available. These will provide service over a variety of geographical coverage areas at various speeds and at a variety of price levels. Each network will serve a niche, but none will meet all needs. Accordingly, mobile computer users will need to access multiple networks in order to meet their needs.

Among the wide area wireless networks available today are: ARDIS, RAM Mobile Data (Mobitex), and cellular digital packet data (CDPD). ARDIS was introduced in the United States in 1983, and RAM Mobile Data began service in the US in 1991. CDPD deployment began in 1994. These wide area wireless networks offer bit rates up to 19.2 kbps.

Wireless local area networks (wireless LANs), on the other hand, offer wireless data service at much higher speeds but over far more limited coverage areas. Intended for in-building use, wireless LANs operate at bit rates up to a few Mbps. A variety of products are on the market which use radio frequencies ranging from 900 MHz to infrared frequencies. Their cell radii are measured in feet rather than miles.

An intermediate service has recently been announced by Metricom. This new 900 MHz service, initially operating in the San Francisco Bay area, uses small "toaster size" base stations that mount directly to utility poles. Cells are approximately 400 meters in radius, and 100 kbps service is provided.

Other intermediate speed networks using the new 2 GHz spectrum recently allocated to personal communication service by the Federal Communications Commission may be introduced within a few years. The speeds and cell radii used in these networks will be decided by the carriers, who will in turn make such decisions based on technological and market factors.

One can observe a speed-distance tradeoff in the wireless networks described. As the cell radii and coverage areas increase, the data rates decrease. In general one can expect that a variety of wireless data networks, with various coverage areas and bit rates, will be available to users. Further, it is likely that high bit rate services will be available only in limited geographical coverage areas, while lower bit rate services will be available over larger areas. A mobile terminal, in order to take advantage of the highest speed network available, will need to be capable of working with any of two, three, or even more wireless networks. Software which allows automatic connection to the most appropriate network, automatic handover to a new network when appropriate, and automatic adjustment to the network's speed will be a very valuable tool to users of mobile computers. Development of such software is the focus of our work.

III.   The Wireless Infrastructure

In order to support this work, we are developing a wireless infrastructure that will allow Carnegie Mellon researchers and others to use mobile computers to gain access to wireless networks both on-campus and off-campus in the greater Pittsburgh area. The wireless infrastructure will be composed of two wireless networks initially, each of which will provide access to our wired campus network.

Wireless service in the wider area will be supported through the use of cellular digital packet date (CDPD) service, which has been introduced by Bell Atlantic Mobile Systems in some parts of its service area. The service is available throughout metropolitan Pittsburgh and operates at a speed of 19.2 kbps.

To complement the CDPD service and provide a much higher speed service on a part of the campus, we are building a high speed wireless infrastructure. This system will use wireless LAN technology to provide high speed (2 Mbps). Some of the high speed wireless infrastructure is expected to be operational by fall 1995.

With both the CDPD and wireless LAN networks operational, properly equipped mobile computers will be able to use both networks. We are currently using laptop computers equipped with Personal Computer Memory Card International Association (PCMCIA) wireless LAN cards and on-board CDPD modems, and we expect to continue their use in the near term. With both wireless networks operational, we will have established a testbed on which the above described software can be developed. With the software operational, mobile computer users will be able to seamlessly move between the wireless LAN and CDPD networks. The infrastructure that has been described is intended to be used both by researchers developing the software and by the faculty, staff and students of the Carnegie Mellon community.

IV.   Cellular Digital Packet Data (CDPD)

CDPD supports the Internet Protocol (IP); thus CDPD service can be easily integrated into an existing IP network by making standard connections between the wired network's router and the CDPD carrier's router. If cellular carriers deploy CDPD by adding it to all of their cell sites, it will offer very wide coverage 19.2 kbps service.

CDPD utilizes existing cellular telephone networks to transmit data. It uses idle channels in existing analog (AMPS) cellular systems to provide a connectionless service. When a channel is not being used for voice traffic, it can be used to send and receive packets from a CDPD-equipped mobile station. Cellular channels can be shared between voice and data service in this way, or the cellular carrier can remove some channels from voice service and dedicate them exclusively to data service.

Thus, CDPD modems "sniff out" available channels on the existing cellular voice network to transmit packets of data. Data from the transmitting device is segmented, encrypted, and formatted into 138 byte frames by the CDPD subscriber device. These frames are sent over one of the 30 kHz channels in the cellular network using a protocol called digital sense multiple access with collision detection (DSMA/CD). DSMA/CD is similar to the carrier sense multiple access with collision detection (CSMA/CD) protocol used in IEEE 802.3 local area networks. The transmitting device listens to find an empty channel; if there is one, it transmits a frame immediately. If not, it searches for a free channel and sends data when a channel has been found.

CDPD operates at layers 1 and 2 of the OSI model and provides service to either of two layer 3 protocols: IP or the ISO Connectionless Network Protocol (CLNP). While the service operates at 19.2 kbps, actual throughput is well below that figure because of the overhead bits required and because of the DSMA/CD contention scheme that is used to gain access to a channel.

A CDPD network serves a mobile device (mobile end system, or M-ES) through a radio link to a nearby cell site equipped with a mobile data base station (MDBS). The MDBS, in turn, communicates with a mobile data intermediate station (MD-IS), which routes packets to and from MDBSs and is typically located at the mobile telephone switching office. Each MD-IS serves a number of MDBSs. Where communications between the M-ES and MDBS and between the MDBS and MD-IS are governed by the layer 1 and layer 2 protocols contained in the CDPD specification, the M-ES to MD-IS communication also uses either IP or CLNP at layer 3. The MD-IS can communicate by wire with an intermediate system (IS), e.g., a router, or a fixed end system (F-ES), e.g.,. a fixed host computer, using IP or CLNP. The ability to support IP is a very attractive feature to many organizations because IP networks are so common.

To use CDPD, one must order the service from a cellular company, establish a wired link to ones own data network or host computer, and buy the required CDPD modem equipment. We selected the Pacific Communications Sciences Inc. (PCSI) Ubiquity 1000 modem because it is the only currently available CDPD modem capable of on-board installation. It can be installed in the floppy drive slot of an IBM Thinkpad 750 series, 755C, or 755CS machine. For all other

Thinkpad models, PCSI should be consulted to verify that the Ubiquity 1000 hardware is supported.

Since this unit must be able to access the cellular network, it contains a 800 MHz transceiver, and it can also be used as a wired phone, a wired modem, a cellular phone, and an analog cellular modem. The unit answers the Hayes AT command set, and PCSI has extended the command set to add CDPD and cellular telephone features along with the standard commands. Since the Ubiquity 1000 contains an analog cellular modem, it can also be used to send data over cellular networks where CDPD is not yet available.

The CDPD network can provide communication between M-ESs without any use of wired data networks. An MD-IS can accomplish this by routing packets between appropriate MDBSs. To provide connections to a fixed host computer or to a wired data network, however, it is necessary to connect a router associated with the CDPD network to the fixed host or to a router which is part of the wired network.

Since we wanted to provide mobiles with access to all of Carnegie Mellon's campus network, we needed to effect a router to router (IS to IS) connection. We purchased a Cisco router specifically for this purpose, and linked it to a Bell Atlantic Mobile Wellfleet router which is part of its CDPD network. The resulting connection provided the needed link between networks. We chose to do this using a 56 kbps Frame Relay connection. Although the 56 kbps service will serve Carnegie Mellon's immediate needs, we plan to upgrade to a T1 link as traffic requires. DSU/CSUs were used to interface each router to a Frame Relay permanent virtual circuit.

When interconnecting two IP networks, it is necessary to make IP address assignments which are acceptable to the operators of both networks. These assignments, which were negotiated between Carnegie Mellon and Bell Atlantic based on the IP addresses available to each, were the most challenging part of the CDPD implementation.

V. High Speed Infrastructure

To complement this CDPD service and provide an infrastructure that will provide much higher speed service, we are presently building a high speed (2 Mbps) wireless network on the Carnegie Mellon campus. We are using wireless LAN technology to provide this service.

Wireless LANs were originally intended to allow LAN connections where premise wiring systems are inadequate to support conventional wired LANs. Now, because the equipment is available in PCMCIA form factor, wireless LANs have come to be identified with mobility. Wireless LANs can be used to provide service to mobile computers throughout a building or throughout a campus.

Many wireless LANs operate in the unlicensed industrial, scientific and medical (ISM) bands at 900 MHz, 2.4 GHz, and 5.7 GHz. Spread spectrum techniques are used in these products, typically direct sequence spread spectrum at 915 MHz or frequency hopping spread spectrum at 2.4 GHz or 5.7 GHz. Other wireless LAN products are available which operate in the (licensed) 18 GHz band and at infrared frequencies.

Wireless LANs operate at speeds up to a few Mbps and have radio ranges from 50 feet to over 1000 feet, depending on the specific product and the environment in which it operates. They can be interfaced to IEEE 802.3 and 802.5 wired LANs, depending on the specific product. A new standard, IEEE 802.11, is being developed to allow interoperability between wireless LANs.

A wireless LAN usually includes both access points (APs) and network adapters (NAs). A NA is a unit, typically available in PCMCIA, that is installed in a mobile computer and gives it access to the wireless LAN. The NA includes a transmitter, receiver, antenna, and hardware that provides a data interface to the mobile computer. The AP is a unit which is mounted in a fixed position and is connected to a wired LAN. The AP, which includes transmitter, receiver, antenna, and bridge, allows NA-equipped mobile computers to communicate with the wired LAN. The bridge which is part of the AP routes packets to the wired network or to mobiles within the its coverage area, as appropriate.

Some wireless LANs allow "roaming," the ability for mobile computers to continue to receive service as they move from the coverage area of one AP to the coverage area of another. In order for this service to be provided, it is necessary that the wireless LAN system know in which AP coverage area each mobile is located. Further, the tables of the bridges contained in each AP must be updated as mobiles move from one AP coverage area to another.

In wireless LANs, direct peer-to-peer (mobile-to-mobile) communication can be provided in one of two ways. In some wireless LANs it is possible for a mobile to communicate directly with another mobile. In other wireless LANs, two mobiles, even though they are both within the coverage area of the same AP, can communicate only by having their transmissions relayed by the AP.

In laying out a multiple AP wireless LAN installation, one must take care to assure that adequate radio coverage will be provided throughout the service area by carefully locating the APs. There is an additional issue in a wireless LAN in which peer-to-peer service is provided by direct communication between the mobiles. In this case one must reduce the coverage area of each AP in a way that will insure that any two mobiles within the coverage area will be within radio range of each other. This reduction of coverage areas will mean that more APs will be needed to cover a given service area. Such reduction in coverage area is not necessary where peer-to-peer service is provided through an AP.

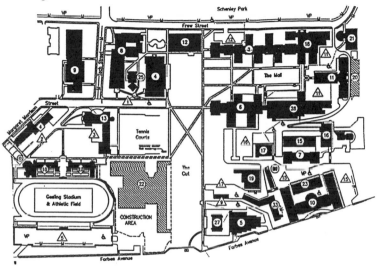

Candidate Building	Number		Candidate Building	Number
Wean Hall	28		Cyert Hall	5
Hamburg Hall	10		Baker Hall	2
Porter Hall	18		Doherty Hall	6
GSIA	8		Hamerschlag Hall	11
			Scaife Hall	21

Carnegie Mellon University Core Campus
Figure 1

We plan to install wireless LAN equipment in several buildings on campus. These buildings house the Carnegie Institute of Technology (the engineering school), the school of Computer Science, the Graduate School of Industrial Administration (the business school), the Information Networking Institute, the Computing Services Department, and other academic and administrative units. Figure 1 shows a map of the core campus. The buildings scheduled for wireless LAN service are listed, along with the numbers which identify these buildings.

We have completed a preliminary design and estimate that approximately 200 APs will be required to cover these buildings and nearby outside areas These will be mounted high on walls inside buildings and on the outside walls of buildings to provide outside coverage. The units will be powered by the campus 110 VAC system.

Each of the APs will be connected, using a IEEE 802.3 10BaseT interface, to the existing campus wired data network. Laptop computers will be outfitted with network adapters (PCMCIA form factor) so that they can utilize the on-campus wireless infrastructure. The same computers are outfitted with CDPD modems so that the computers will have access to both the on-campus wireless network and also the Pittsburgh area CDPD network.

VI.   Acknowledgments

We wish to thank John Leong for the ideas he has contributed. The work reported here is supported by Carnegie Mellon's Information Networking Institute, the National Science Foundation, Bell Atlantic Mobile, and Bell Communications Research.

A CDPD Performance Model for an
Intelligent Transportation System Architecture Evaluation[1]

Steven P. Arnold, Russell W. Taylor, and Mark A. Wallace[2]

Loral Federal Systems
Manassas, VA

Loral's effort to develop an architecture for a nationwide integrated Intelligent Transportation System required an evaluatory communications design. For the wireless component, the Loral team focused on CDPD. This paper details the model that was developed using the OPNET™[3] tool and the simulations that were run to predict performance for an urban scenario in the year 2012. The results indicate that current CDPD technology can handle the required loads, which are primarily associated with route guidance.

1. Introduction

The United States Department of Transportation (US DOT) sponsored four teams in a Phase I effort to develop a national architecture for an Intelligent Transportation System (ITS). The goals of the program were to meet the national needs embodied in 28 "user services" while also defining a system that was open, expandable, and capable of nationwide interoperability. This Phase I effort concluded in November of 1994, and two of the four teams, Loral and Rockwell, are now working on a Phase II effort.

The ITS Architecture defines various subsystems, message sets that are communicated between the subsystems, and message rates for these communications. When this information is combined with a scenario describing a transportation system (numbers and types of vehicles, types of equipment, layout of the road network, etc.) it becomes possible to perform a data loading analysis. This data loading analysis can then be used to estimate the requirements for different communications links in the Architecture.

The Architecture is not intended to dictate which technologies are to be used in deployment; rather it should provide a definition of the requirements to ensure interoperability and openness. However, as part of the Phase I effort, an "evaluatory design" was developed, to demonstrate the feasibility of implementing the various Architectures. In the Loral Architecture, one of the most critical communications links is the wireless one between the vehicles and the various potential providers of information. To satisfy the requirements of a national architecture, the Loral team focused their evaluatory design of the vehicle–infrastructure communications link on an analysis of Cellular Digital Packet Data (CDPD) as the technology of choice. This selection was based on the expected availability of CDPD, the presence of an open standard, and the potential for near–total coverage of the US population (anticipated at 90% by 2002 [Am94]).

Simulation was included in Phase I for both communications and traffic. A scenario called "Urbansville", corresponding roughly to the greater Detroit area, was provided to the teams as a common road network and set of assumptions. The Loral team produced a detailed model of a single CDPD cell and transceiver, and then applied the Urbansville parameters to determine the likely scaling,

1. The authors would like to acknowledge support for this work from Loral Federal Systems Group and the Federal Highway Administration under contract DTFH61–93–C–00024. Some of the results in this paper have been previously released as part of the aforementioned contract.

2. Mr. Wallace is now affiliated with Alcatel Data Networks.

3. OPNET and Transceiver Pipeline are trademarks of MIL 3 Inc.

based on the model, up to a scenario depicting a 2012 case where 50% of the vehicles are equipped with CDPD capability [Lo94a].

The objective was to use the CDPD model in a simulation to demonstrate adequate capacity to support an urban scenario traffic load in the year 2012. The modeling focus was on the vehicle–infrastructure link, and the simulation was used to analyze both the CDPD forward and reverse channels for: (1) Link capacity, (2) Throughput at selected interfaces, (3) Information latency for specific message types, (4) Maximum data transfer rates (growth capacity), (5) Channel utilization (number of vehicles/cell), (6) System capacity, and (7) Estimate of the air channel bandwidth requirement. The approach we used emphasized detailed modeling of the protocol and then used this model in simulations with relatively small numbers of users, extrapolating the results to larger systems.

2. Simulation Approach

This section outlines the steps followed in the Architecture communication system definition, development, and evaluation. These steps were iteratively repeated throughout the phases of the spiral design methodology employed in refining the ITS Architecture. For Phase I, average and peak data loads on communications links were estimated, and then a CDPD model was developed. Figure 1 shows the process for data loading analysis and communication networks simulation.

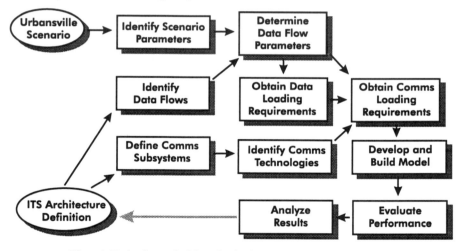

Figure 1. Evaluation methodology for the ITS communications architecture.

The steps involved in defining the Architecture are outside the scope of this paper. However, it is worth mentioning the three main components of the Architecture: the logical architecture, the physical architecture, and the data dictionary. The *logical architecture* [Lo94b] contained more than 1000 data flows needed to satisfy the user service requirements. The messages associated with each data flow were contained in the *data dictionary* [Lo94b]. These data flows were aggregated together in the *physical architecture* [Lo94c] via the process of assigning different logical functions to physical subsystems. The communications modeling and simulation were a critical part of the iterative process of refining the Architecture: as the performance was analyzed, changes in communications requirements could be achieved by moving functions between physical subsystems. Ultimately, Loral's final Phase I Architecture required a delicate balancing of these performance issues against implications for legal, cost, and a host of other issues.

Given an Architecture definition and the scenario parameters, the static data loading requirements were determined. The message length and frequency estimations were used to produce data loading estimates that indicated the bandwidth requirements of each flow. As a result of aggregating the bandwidth requirements of data flows that had the same source and destination subsystems, the communication bandwidth requirements of the interconnection network was obtained.

After identifying the communication technologies to be used, we developed and built the communication models for the proposed communication architecture. For the CDPD modeling, the link of interest was primarily that between the *vehicle* subsystem and the *independent service provider* (ISP) subsystem. While data loading analysis was a bottoms–up effort, model design was a top–down process. Each layer of the network hierarchy was divided into subsystems with link, message, and node descriptions associated with each layer. The protocols and algorithms were also modeled in appropriate detail. Critical parameters were identified and assumptions were noted as part of the model development. Then the model was verified, calibrated and validated before the communication system evaluation began, via either simulation or rigorous analysis.

The final steps were to execute the simulation model and perform analysis. The performance measures were defined and critical parameters were identified. Various workload scenarios were executed to provide data for the evaluation of the communication systems performance. The analysis then uncovered the strong and weak points of the chosen physical and logical architecture. The analysis also provided trade–off studies for different technologies, message descriptions, protocols, etc.

The simulation had two distinct levels of focus. One level of focus was on capacity. The system capacity is a system– or macro–level of measurement. The second level of focus was at the lowest– or "micro"–level of measurement. An example of this would be the measurement of the latency for an individual vehicle sending a particular message such as "route request". The combination of these two aspects then gave a fairly complete picture of the performance and implications of using CDPD for this aspect of ITS.

The lowest level simulation was necessary to understanding the system from an individual users perspective. The approach allowed for the observation of the impact of the layers of protocol at the individual device level. The approach allows the imposition of worst case conditions on individual users and then close scrutiny of the performance of the CDPD protocol and its interaction with the other layers. Insight into the system performance, as perceived by individual users, is very relevant when the architecture under study contains time dependencies.

The simulation results captured consisted of both system performance indicators averaged over the entire simulation time and sampled values that represent averages over a few seconds. The system performance averages are useful in judging the overall utilization of the communication channels, but can easily hide system loading problems that occur for a short duration. When considering a ITS system, instantaneous peaks in bandwidth usage are likely (during a traffic incident, as an example). Therefore, the ITS Architecture and the proposed wireless solution were specifically tested during the occurrence of these sudden peak demands.

3. Developing the CDPD Model

Our goals for modeling the vehicle–infrastructure CDPD link shaped the modeling structure and focus. The model was developed to analyze the impact of an ITS system on a cellular network and to show the adequacy of the CDPD protocol for an evaluatory design. The primary goal of the simulation was to study the performance of the physical layer protocol; the upper protocol layers were modeled to the extent necessary to accurately depict their influence on the CDPD link. As with any modeling activity, there were many assumptions made in developing this CDPD model; we will discuss the basic ones in the following section.

3.1 Modeling Assumptions

The CDPD specification was not modeled in its entirety, rather the portions of the specification that significantly impacted the physical channel were implemented. The CDPD specification encompasses many functions that use bandwidth, but only the major influences were considered. In particular the layer protocol's contribution to overhead and latency was explicitly modeled, while some of the layer management services were not. An example MDLP layer management service that was not modeled was the sleep function used to support power conservation, since it does not significantly impact our area of study.

A fundamental assumption was that the layers above the CDPD physical layer could maintain sufficient data rates to handle the physical layer's requirements. Thus, while we fully modeled all delays in the various layers, a data handling bottleneck could only occur at the physical layer.

Another basic assumption was that the channels in this model were dedicated to CDPD. This was mainly an assumption to simplify the models and, given that CDPD channels have been dedicated in some cities [Am94], an accurate reflection of reality. The influence of normal AMPS traffic and cochannel interference was not modeled in the radio link. The model established the increased load imposed on the system by ITS, so the existing traffic load was not deemed a relevant parameter.

In performing the simulation under different demand conditions it was necessary to define the point of failure of the CDPD communication system. A transmission failing at the MAC level (due to a maximum number of attempts) will be be resent by the next layer (MDLP). The MAC layer now can start trying to send the message again. If the TCP layer does not get a response in a specific amount of time it will attempt to send the message again. This scenario is true at all layers of the protocol stack. If the application is willing to wait long enough, it will have a very high success rate of getting a message transmitted and acknowledged. Of course, applications exchanging information need to perform tasks at a minimum desired speed, so some definition of inadequate performance was needed. For this simulation, a transmission failure was defined to be a MAC layer attempt timeout on 3% of attempts.

To summarize:

- Model CDPD system parameters are specification default values.
- Upper level protocol layers transfer data faster than CDPD layers.
- Data sizes and frequencies based on Data Loading and Urbansville parameters.
- Transmission failure was defined to be a MAC layer attempt timeout on 3% of attempts.

3.2 Model Layers

There were two types of nodes in the independent service provider–to–vehicle (*ISP–Vehicle*) model. The ISP node was a stationary node that communicates with many mobile nodes representing the vehicles. Each node contained a transmitter, receiver, and a full CDPD protocol stack. The ISP node transmitted information using the forward channel and received information using the reverse channel. The forward channel was a contention–less broadcast channel carrying transmission from the ISP only, hence capacity of the forward channel was high. This is important for ITS applications since the largest amount of information (routes) would be sent on this channel. The vehicles shared the reverse channel for transmitting information and received information on the forward channel.

The CDPD specification itself was the basis for the ISP–Vehicle model. The model timings and functions were taken directly from the specification. The Physical and Media access layers were modeled and validated using the specification. Since the focus of the model was to determine channel capacity, the channel access methods and protocols were explicitly modeled. The system simulations involved the interaction of many CDPD service users. To validate that the model was behaving realistically, performance data was obtained from Ameritech [Am94] to calibrate the system model. The basic CDPD layers are shown in Figure 2.

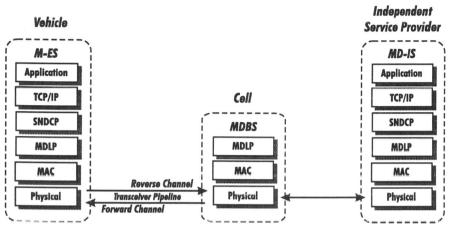

Figure 2. CDPD protocol stack and model components.

The OPNET™ model was built by modeling each protocol layer as described in the CDPD specification. This includes modeling the state machines, timers, and command formats. The following outlines some of the model features:

- *Physical layer* explicitly modeled

- *MAC data link layer* explicitly modeled: (1) Full duplex; (2) Data encapsulation, framing, and synchronization; (3) Error detection; (4) Medium allocation; (5) Contention resolution, and (6) A defer algorithm with maximum attempt timeout. The state diagram for the MAC layer transmit function is shown in Figure 3, as an example of the state diagrams produced for all the layers in the model.

- *MDLP Layer*: (1) Complete data link services; (2) Complete frame and command formatting; (3) Point–to–point information transfer (both acknowledged and unacknowledged); (4) Broadcast information transfer; (5) Complete protocol procedures; (6) Command sequence timeout; (7) SNDCP modeled to support TCP/IP network layer; (8) TCP/IP layers (as provided in the OPNET™ library), and (9) Application layers modeled as message source and sinks with message sizes and frequencies defined by the architecture.

- *The wireless portion* of the CDPD model was simulated by the radio line Transceiver Pipeline™ available in OPNET™. The fourteen stage pipeline provides mechanisms for link closures, channel matching, transmitting antenna gains, propagation delays, receiving antenna gains, received power, background noise, interference noise, signal to noise ratios, bit error rates, and error allocation and correction [Op93]. This granularity and flexibility allowed for detailed analysis of the CDPD radio link from the vehicles to the cell sites.

- Model timings and capacities verified against the CDPD specification.

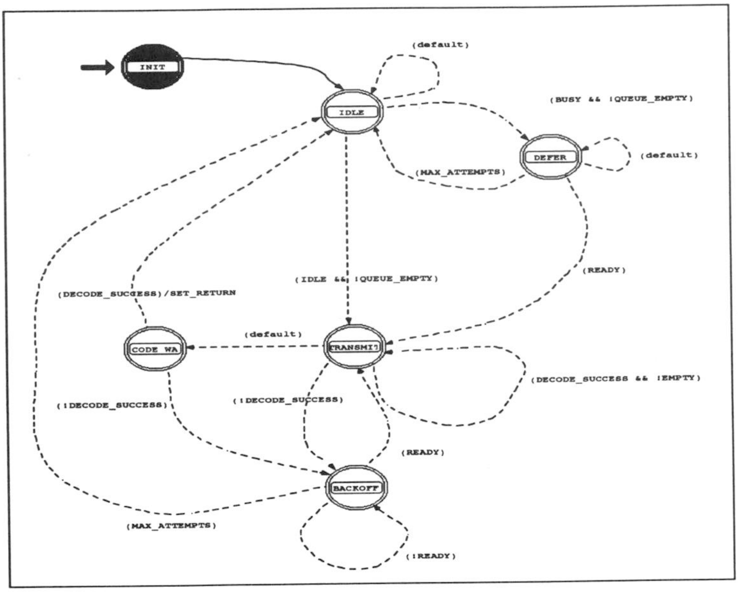

Figure 3. State diagram for the CDPD MAC layer transmit function.

Once the model was constructed, the data to be exchanged was defined. The average message frequency and size for the system simulation were derived from from the data loading analysis, which represented the results of a steady–state analysis. These averages represent a value determined over a long period of time. To prove our communication system was adequate we had to take into account the minute–to–minute variations that could occur due to statistical variability within a chosen time interval. The average message size and message frequency from data loading were taken as the mean value for use in the appropriate distribution function (Poisson or Periodic).

4. Simulation Results

There were three main steps in our communication analysis process: the data loading analysis, the CDPD cell model, and the extrapolation to a full urban region. During the data loading analysis the effects of message size and frequency variations were examined. The next step was to take these steady–state results and apply the message characteristics to our CDPD model. The message size and frequency values were used as the mean values for the distribution functions used in simulating a representative model of the CDPD system. The stochastic value from the distribution represented a time varying load that allowed the study of the system as a dynamic environment. And finally, the results of the model simulation, in conjunction with Urbansville parameters, were used to project full size system performance and to draw conclusions about the system capacity.

4.1 Simulation Parameters

The CDPD simulation centered around a fictitious urban area named Urbansville with a road network similar to metropolitan Detroit. The parameters for Urbansville include the following: a population of 3,788,000, 800 square miles in area; 1282 miles of freeways and arterial roads; 2,159,000 automobiles, and an average of 329,000 on the roads during peak traffic hours. For the year 2012 it was assumed that 50% of the vehicles were equipped with ITS devices capable of receiving route guidance, giving a total of 164,500 equipped vehicles.

The CDPD simulation focused on two loading levels: peak and average. *Peak* captured our worst-case assumptions, such as traffic loads during "rush hours", which produce the heaviest load on the road infrastructure as well as the cellular communications infrastructure. *Average* captured the assumptions for average usage over a typical 24 hour period. Both loading scenarios were modeled.

Many different data messages are sent between a vehicle and an independent service provider (ISP) using CDPD. However, the CDPD model focused on route guidance data, which is typically the largest data message sent to a vehicle, according to the Loral ITS Architecture. The message size was consciously conservative and did not assume any data compression. Many other ITS related data messages (for example probe data, where a vehicle reports on current traffic conditions) were also simultaneously modeled over the CDPD network.

The results from the Loral ITS analytical data loading analysis were used to determine the frequency that route guidance messages were sent from the ISP to the vehicle. The following parameters were used [Lo94a]:

- Vehicle to ISP peak loading is 4.1 bits/sec/vehicle
- Vehicle to ISP average loading is 4.0 bits/sec/vehicle
- ISP to Vehicle peak loading is 52.0 bits/sec/vehicle
- ISP to Vehicle average loading is 32.0 bits/sec/vehicle

These figures derive from the combination of hundreds of parameters for message size, frequency of transmission, trip duration, route length, and a number of other aspects. All of these parameters are recorded in the Loral Data Loading Requirements [Lo4a] and Logical and Physical Architecture documents [Lo94b, Lo94c]. These assumptions require further scrutiny, which will occur during Phase II.

The cellular infrastructure for the CDPD model was based in the central business district of Urbansville, the most populous location. Urbansville was assumed to be comprised of 120 degree sectorized cell sites. The spacing between these cell sites was governed by the ratio of the distance (D) between same–frequency cell sites and the radius (R) of the cells. It was assumed that D/R ≥ 4.6 [Le89]. A typical cell in the central business district of Urbansville would have a diameter of approximately 2.0 miles [Am94].

The transmission delays were recorded under the constraint that a transmission failure is defined to be a MAC layer attempt timeout on 3% of attempts. Transmission delay is defined to be the time required to send a message between the ISP application and the vehicle application. Data throughput was analyzed by attempting to send information at particular rates and observing the actual transmission rates under the failure constraint mentioned previously. An example of observing transmission rates would be for the aggregate users to attempt to send data at 10,000 bits/sec and observing that the channel throughput is 8000 bits/sec. This difference in attempted vs. observed throughput is caused by protocol overhead, channel contention, and transmission failures. These factors result in transmission delays.

The some of the simulation results are in the figures that follow. Terminologically, the average throughput graphs show a running average for the simulation. All other "averages" were taken over a moving window of time. And the instantaneous plots show individual simulation events. All three plots are important, yielding respectively steady–state, short term average, and instantaneous peak measures. A few example results figures are shown in this paper.

The top graph in Figure 4 shows the overall transmission latency between the ISP application and the Vehicle application. This delay essentially encompasses the protocol stack delays and the CDPD link delays. The second graph in Figure 4 shows the average throughput in bits/sec on the reverse channel under conditions of maximum channel utilization. The slope at the beginning of the average throughput graph represents the simulation reaching an equilibrium. Maximum channel utilization is recognized by the number of transmission failures occurring as discussed earlier. This failure behavior is shown in Figure 5, which depicts the instantaneous fraction of transmission failures over the simulated interval.

The following outline summarizes the simulation results that were then used in the analysis:

• Results include all of the TCP/IP protocol and the CDPD protocol.

• The best case results for the forward channel (downlink) was an average of 15,000 bits/sec when an 8000 bit message was sent. A route was roughly 8000 bits. The instantaneous queueing backup at the CDPD MAC level became significant at this rate. The queuing backup was sometimes 5 routes deep. Note that the interarrival rate was a distribution so the instantaneous MAC queuing varied. This forward channel performance is shown in Figure 6.

• The reverse channel (uplink) has an average bits/sec of 7500 bits/sec when the bandwidth of 8000 bits/sec is attempted by the mobile units. At this point the transmission failure rate begins to exceed 3%. The mobile units are sending 1000 bit messages. (route request). Simulation showed that 9000 bits/sec average was the highest throughput for the uplink channel. The small packet size being sent emphasized the impact of the overhead. The vehicles also sent small data sizes when probe data was transmitted. These messages were smaller than the route requests, but more frequent. The uplink behavior is captured in Figure 4 and Figure 5.

• Additional results not captured in the figures: application to application communication delays are usually under 2 seconds; the MAC reverse channel average delay introduced is under 0.5 seconds, with peaks slightly over 1 second; and the MDLP average delay introduced is under 1 second, with peaks slightly over 5 seconds.

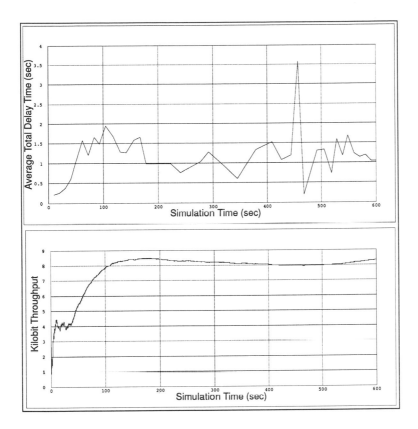

Figure 4. Reverse Channel Transmission Latency (top) and Average Throughput (bottom).

4.2 Simulation Analysis and Extrapolation

The analysis method combined the simulation results with analytical information to reach conclusions on system capacity. The analytical parameters and the simulation results were extrapolated to estimate the CDPD communications loading on the cellular network in Urbansville. It was assumed that about 250 cells would be needed for Urbansville by the year 2012.

Assuming a uniform distribution, 164,500 CDPD–equipped vehicles over 250 cells yields 658 vehicles/cell. We can then extrapolate:

- Vehicle to ISP bandwidth peak is 2.6 Kbits/sec for 658 vehicles

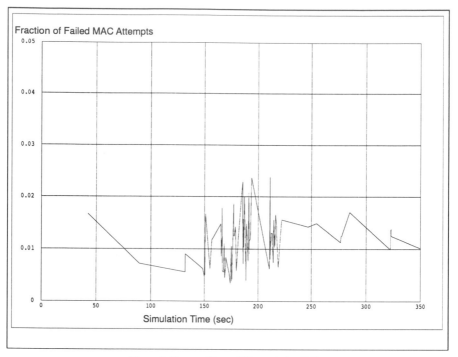

Figure 5. Reverse Channel Transmission Failures

- Vehicle to ISP bandwidth average is 2.6 Kbits/sec for 658 vehicles
- ISP to Vehicle bandwidth peak is 34.2 Kbits/sec for 658 vehicles
- ISP to Vehicle bandwidth average is 21.1 Kbits/sec for 658 vehicles

The CDPD channel requirements for the Loral ITS Architecture are as follows:

- Reverse Channels: One (1) channel is needed for peak periods.
  Less than one channel is needed for average usage.

- Forward Channels: Around five (5) channels are needed for peak periods.
  Around three (3) channel are needed for average usage.

- CDPD does not significantly effect information latency.

Most of the communication delay will probably occur at the data switch leaving the cell and in the connection from the mobile data information system to the ISP. The cell was assumed to typically be connected to the MD–IS using one time slot of a T1 link (64 Kbits/sec). The MD–IS was connected to a Local Area Network (LAN) at the Mobile Telephone Switching Office (MTSO) and the LAN was connected to a packet switched network.

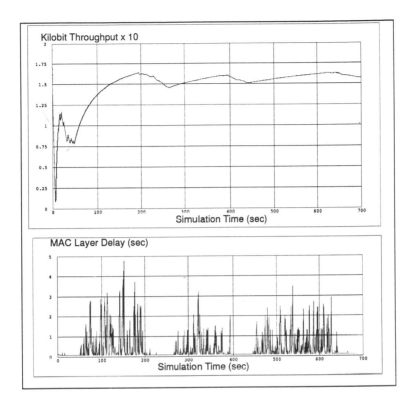

Figure 6. Forward Channel Throughput (top) and instantaneous MAC Delay (bottom).

5. Conclusions

The results of the OPNET ™ simulation of the Urbansville scenario parameters indicate that the communication requirements placed on the CDPD infrastructure by the Loral architecture are supportable. Assuming that there are *two vendors* in each wireless market providing CDPD, *each* vendor has to have 3 forward channels available and 1 reverse channel available for peak usage. Since the channels are paired, the channel requirement for peak loads is three channels per cell. The currently implemented CDPD systems can adequately accommodate this level of IVHS traffic. It should be noted that the analysis was done in the 20 year time frame with worst case Urbansville scenario values.

The forward channel is the potential bottleneck in the Loral ITS Architecture. Several things should be considered when sizing the route communication load. In response to an incident, partial routes could be sent. These partial routes could be a modification to the existing route and therefore be much smaller than the route sizes assumed in our simulations. Also, this analysis does not account for broadcasting routes to more than one vehicle. This would greatly reduce the requirements. This idea involves grouping vehicles with destinations in a common localized geographic area, or grouping vehicles that are to be affected by an incident. For instance, when an incident occurs on an interstate highway, the information concerning an appropriate detour route would be common for many of the vehicles approaching the incident location. Refinement of the architecture and the actual preliminary design of the communication and messaging system would probably result in a lower predicted usage of the CDPD infrastructure.

TCP/IP places a significant overhead load on the system both in the number of messages sent and in the overall message size. The size of the TCP/IP overhead is typically larger than the messages being sent on the interface. For messaging, other protocols such as UDP may be a better solution. Also, the protocol layers of CDPD have well defined and proven methods of ensuring reliable data transfer, so the added value, if any, of TCP/IP is worth evaluating further.

Two issues for future work are to better tie the traffic simulations to the communications simulation and to estimate the non–ITS CDPD background traffic. The former would allow us to uncover any inadequacies resulting from the assumption of a uniform traffic distribution; without doubt the communications requirements associated with a traffic backup at an incident will deviate from our simplified assumptions. The latter issue, estimating background CDPD usage, is required to test the conclusion that there is adequate bandwidth available to support ITS. Given competition from other, more lucrative uses of CDPD, it is possible that ITS–based applications might see a degradation in performance or availability.

Future digital data transmission systems are likely to be much better than CDPD. CDPD represents an adaptation of an existing system and thus has overhead and limitations associated with its compatibility with AMPS. The future communications systems will probably be an implementation of Code Division Multiple Access (CDMA) or Time Division Multiple Access (TDMA) protocols. These future systems should even more easily absorb the ITS communication requirements than has been shown for the CDPD system.

6. References

[Am94a]   Ameritech Mobile, personal correspondence, May 1994.

[Ce93]   *"Cellular Digital Packet Data System Specification"*, Release 1.0, July 1993.

[Le89]   Lee, William C. Y., *"Mobile Cellular Telecommunications Systems"*, 1989, 449 pp.

[Lo94a]   *"Analysis of Data Loading Requirements"*, prepared for FHWA by Loral Federal Systems, document no. FHWA–LFS–94–038, October 1994, 110 pp.

[Lo94b]   *"IVHS Logical Architecture Document"*, prepared for FHWA by Loral Federal Systems, document no. FHWA–LFS–94–035, October 1994, 507 pp.

[Lo94c]   *"IVHS Physical Architecture Document"*, prepared for FHWA by Loral Federal Systems, document no. FHWA–LFS–94–036, October 1994, 292 pp.

[Op93]   *"OPNET Modeling Manual 2.0"*, MIL 3 Inc., November 1993.

Capacity of Channel Hopping Channel Stream On Cellular Digital Packet Data (CDPD)

Jay M. Jacobsmeyer

Pericle Communications Company
P.O. Box 50378
Colorado Springs, CO 80949
(719) 548-1040
E-Mail: 73251.1500@compuserve.com

Abstract—*Because voice cellular radio systems must maintain low blocking probabilities, almost 20% of cellular channel capacity is unused, even during the busy hour. A packet radio system can capture this unused capacity by hopping between idle voice channels, working independently of the voice base station and the cellular telephone switch. To optimize performance, the mobile data base station should select the idle channel with the least co-channel interference. By doing so, the base station will maximize the carrier-to-interference ratio at the mobile terminal and maximize throughput when automatic repeat-request protocols are used. For example, if the offered voice traffic at a 54-channel base station is 22 Erlangs, the data base station can achieve a throughput improvement of a factor of 2 and a coverage area increase of 47%.*

1.0 INTRODUCTION

Until recently, the North American cellular radio network was used almost exclusively for voice communications. Today, there is a high demand for data communications on the cellular network. The most straightforward approach to providing data service on the cellular network is to connect a wireline modem (e.g., V.32*bis*) to the audio input of the cellular radio.

Unfortunately, this approach is fraught with problems and performance has been disappointing.

The main problem is the random amplitude and phase modulation caused by multipath fading. Wireline modems use dense signal sets that are very sensitive to amplitude and phase fluctuations. The inability to track amplitude and phase results in high packet error rates. In addition, multipath fading tends to wash out the FM capture effect that is helpful with stationary receivers. Thus, the cellular radio frequency deviation (12 kHz) offers no signal-to-noise ratio benefit for a moving receiver in a multipath environment.

A logical solution to this problem is to employ digital modulation techniques with robust error control and directly modulate the radio frequency carrier. This solution is being fielded as part of the Cellular Digital Packet Data (CDPD) system.

CDPD is designed to provide cost-effective data service over existing cellular radio channels. CDPD operates independently of the voice system, using either dedicated 30 kHz channels or hopping between idle channels. The CDPD channel, called the *channel stream*, is a shared medium employing digital sense multiple access (DSMA) with collision detection (CD). The system uses its own base station, called a mobile data base station (MDBS) and communicates with data mobiles, called mobile end systems (M-ESs). Calls are routed by intermediate systems (ISs), also called *routers*. The system block diagram is shown in Figure 1. CDPD is described in more detail in a specification [4], first published in July 1993. Service was available in several U.S. cities in 1994.

[1]This material is based upon work supported by the National Science Foundation under award number 9461685. This work was supported in part by Steinbrecher Corporation, Burlington, MA.

40

One of the key advantages of CDPD is its ability to exploit unused capacity in the cellular network.

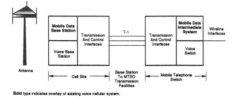

Figure 1 - CDPD System Block Diagram

Because mobile users access base stations at random intervals and remain on the system for random durations, the base station can not use all of its inherent capacity and maintain acceptable blocking probabilities. For example, a 54-channel base station with 2% blocking serves only 43 users on the average. Thus, there is an excess average capacity of 11 channels, or 20%. This principle is shown graphically in Figure 2 where we have plotted the number of busy channels versus time for a multiserver loss system (from simulation).

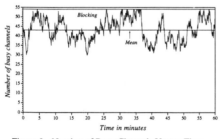

Figure 2 - Number of Busy Channels Versus Time
($m = 54$, $\lambda/\mu = 44$)

CDPD exploits this excess capacity by hopping between idle voice channels. This capacity improvement comes at a price, however. The hopping channel stream adds co-channel interference that ultimately reduces geographical coverage for voice and data users.

The purpose of this paper is to estimate the capacity of the channel stream, determine the impact of the channel stream on voice channel blocking and co-channel interference, and describe some approaches that will optimize hopping performance.

2.0 CHANNEL MODEL

The communications channel is the terrestrial mobile radio channel described in detail in [1]. Multipath fading causes a time-varying, Rayleigh distributed signal amplitude at the mobile radio receiver. The local mean, Y, of the fading signal is lognormally distributed and the mean of Y is proportional to $r - \gamma/10$ where $\gamma/10$ is the propagation exponent. The velocity of the mobile unit relative to the base station causes a doppler spread with maximum value given by

$$f_m = V/\lambda_w \qquad (1)$$

where V is the vehicle velocity and λ_w is the wavelength of the radio carrier. We assume negligible delay spread.

We shall use the 7-cell frequency reuse pattern shown in Figure 3. In a 7-cell reuse system, there are a maximum of 6 co-channel interferers in the first tier of co-channel cells. Because of the larger path distance, interference from the outer tiers of co-channel cells is small relative to the interference from the first tier. We shall neglect adjacent channel interference.

To minimize co-channel interference, cellular operators often employ directional base station antennas. For perfect antennas with 120° beamwidths, the number of co-channel interferers is reduced from 6 to 2. In practice, perfect antennas do not exist and the maximum antenna front-to-back ratio is typically 10 dB [3].

We assume that background noise and co-channel interference are additive, white and Gaussian distributed. All interference sources are uncorrelated. This is a worst-case assumption in the sense that it maximizes the variance of the carrier-to-interference ratio, C/I. Co-channel interference is time-varying due to random channel activity in co-channel cells. Co-channel interference is typically much higher than the background noise, and we shall assume that the ratio of mean interference power from one interferer, I_1, to the noise power spectral density, N_0, is 10 dB. Interference from outer tiers of co-channel cells is considered constant with time and is included in N_0.

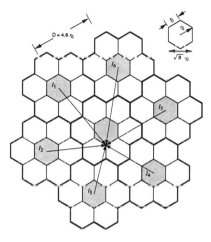

Figure 3 - Co-Channel Interference Sources in a 7-Cell Frequency Reuse Pattern

3.0 MULTIPLE ACCESS MODELS

The population of mobile radio users consists of voice users and data users. Voice users have priority access to the radio resource.

A. Voice Users

The multiple access model for mobile voice users is the *multiserver loss system*, also called the *Erlang loss system* [8]. Voice users arrive at the base station according to a Poisson random process with rate λ. There are m servers (channels), each with independent, identically distributed service times. Arrivals who find all m servers busy are turned away and lost to the system. (They get a busy signal). No queuing of arrivals is allowed. Service times (call durations) are exponentially distributed with mean $1/\mu$. For an m server system, the probability that k servers are busy is given by [8]

$$P_k = \frac{\left(\frac{\lambda}{\mu}\right)^k}{k\,!} \left(\sum_{j=0}^{m} \frac{\left(\frac{\lambda}{\mu}\right)^j}{j\,!} \right)^{-1}$$

(2)

and the mean number of busy servers is

$$\mathrm{E}[K] = \frac{\lambda}{\mu}\,(1 - P_m)$$

(3)

where P_m is the probability that m servers are busy, also known as the *blocking probability*, P_b.

B. Data Users

The multiple access model for mobile data users is the slotted, non-persistent Aloha model [6]. In this paper, we are primarily interested in the availability and potential throughput of the channel stream. We shall not examine data user multiple access performance in any detail.

In general, the channel stream can occupy any one of the m radio frequency channels that is currently idle. The data system operates independently of the voice system by sensing voice activity and hopping between idle channels. When all m channels are required for voice use, the channel stream must hop off the system entirely and data service is unavailable until a voice service is completed and a voice channel becomes idle.

4.0 IMPACTS ON VOICE USERS

The potential impacts of a hopping channel stream on voice users fall into two categories: call blocking and co-channel interference.

A. Blocking

The channel hopping channel stream must not increase voice call blocking. Therefore, the channel stream must hop off the system entirely whenever all m channels are needed for voice use. The maximum availability of the channel stream is thus $1-P_b$ where P_b is the probability of voice channel blocking.

A dedicated data channel would be much easier to implement than a hopping data channel, so, why don't we just dedicate a voice channel to data use? The problem is that dedicated channels increase call blocking for voice users. For example, consider a 54-channel base station with a blocking probability of 2% ($\lambda/\mu = 44$). If we remove three voice channels for data (one per sector), only 51 are left for voice users. Computing (2) for $m = k = 51$ and $\lambda/\mu = 44$, we get a blocking probability of 3.8%, a 90% increase in the number of blocked calls. Another way to look at the blocking problem is to argue that we must maintain a blocking probability standard (e.g., 2%) and by removing a voice channel from service, we remove potential voice customers. For example, during the busy hour, a single voice channel at a single base station will support 48 minutes of air time ($m = 54$, $P_b = 2$%). A hopping channel stream allows the cellular operator to provide this airtime and collect the resulting revenue.

B. Co-Channel Interference

A hopping channel stream adds average interference to co-channel cells that would not be present otherwise. The main source of this co-channel interference is the first tier of co-channel cells, shown in Figure 3. But does this interference have a noticeable effect on voice users? To answer this question, we need a relevant measure of the interference introduced in co-channel cells. The most appropriate measure is the probability that a randomly selected point in the home cell (the center cell in Figure 3) will have a mean C/I of u_{min} dB or greater. In other words, we want the probability that a mobile user has a $C/I \geq u_{min}$ dB, assuming the mobile user's position is uniformly distributed over the cell's area. Typical values of u_{min} are 17 or 18 dB and are determined by the subjective voice quality of the FM radio receiver in Rayleigh fading [2].

In this section, we are interested in the long-term fading effects on C/I and we will therefore ignore Rayleigh fading for now. Define the *coverage area* as the fraction of a circle of radius r_0 for which the C/I is greater than u_{min}.

The interference power, $Z = 10^{I/10}$, is the sum of received powers from up to six sources. We assume that the sum of lognormal random variables is well approximated by a lognormal random variable [10]. Therefore, the interference, I (in dB), is lognormally distributed with mean m_I and variance σ^2_I. The desired signal, C (in dB), is also normally distributed with mean m_C and variance σ^2_C. Expressions for the mean and variance of C and I are derived in the Appendix.

We are interested in the random variable, $U = C - I$. The interferers are separated by 60° in azimuth and it is reasonable to assume that the terrain effects will be uncorrelated, so the independent assumption for interferers is appropriate. The desired signal may be correlated with one or two interferers, but the independent assumption is worst case in the sense that it maximizes the variance of the carrier-to-interference ratio. From [7], we know that the sum of independent normal random variables is also a normal random variable with mean equal to the sum of the means and variance equal to the sum of the variances.

Therefore, the carrier-to-interference ratio, U, is normally distributed with mean $m_C - m_I$ and variance

$$\sigma^2 = \sigma^2_C + \sigma^2_I \qquad (4)$$

In the Appendix, the following expression is derived for the coverage area of a cell in the presence of co-channel interference:

$$P(U > u_{min} \mid w) = 1 -$$

$$\frac{1}{\pi} \int_0^{2\pi} \int_0^1 \Phi\left(\frac{u_{min} - m_U(s, \theta)}{\sigma}\right) s \, ds \, d\theta \quad (5)$$

where w is the number of active co-channel interferers, $w = 1, 2, ..., 6$ for an omnidirectional cell and $w = 1$ or 2 for a directional cell; s is the ratio of the mobile unit's distance from the base station to the cell radius, $s = r/r_0$; $\Phi(x)$ is the unit normal distribution,

$$\Phi(x) = \frac{1}{\sqrt{2\pi}} \int_{-\infty}^x e^{-\frac{y^2}{2}} \, dy$$

and $u(s, \theta)$ is the mean C/I at polar coordinates (s, θ) in the cell,

$$m_U(s, 0) = m_C - m_I$$

Equation (5) can be solved using numerical integration and is plotted in Figure 4 for $\sigma_C = \sigma_{I_1} = 8$ dB and $\gamma = 40$ dB/decade ($\sigma_{I_1} =$ standard deviation for a single interferer).

Note from Figure 4 that worse-case coverage ($w=6$) for $C/I = 17$ dB is 69%, which is far less than the usual design goal of 90%.

Partitioning a cell into 3 sectors with 120° beamwidth antennas will increase the mean C/I. Assuming perfect antennas, only cells 1 and 2 of Figure 3 contribute to co-channel interference in a 3-sector system. From Figure 4 ($w = 2$), we see that this increase in C/I raises the area covered with greater than $u_{min} = 17$ dB to 85%.

Recall that our purpose in deriving (5) is to determine the effect of a hopping channel stream on average co-channel interference and coverage. For a 54-channel base station with 2% blocking, there are 43 busy channels, on the average. In an omnidirectional cell, the channel hopping channel stream adds one busy channel for a total of 44.

Interference in the system is increased by a factor of 44/43 or 0.1 dB. From Figure 4, we see that this added interference reduces our coverage by only 0.3% ($w = 5$, $u_{min} = 17$ dB). In a sectorized cell ($m = 18$, $\lambda/\mu = 14.7$), the increase in co-channel interference is 14.5/13.5 or 0.3 dB, resulting in 0.8% loss in coverage. As new channel streams are added, the average interference will increase, but the maximum increase in interference is 54/43 or 1.0 dB. This equates to a loss in coverage of 2.7%. Whether this loss in coverage is acceptable or not is the decision of cellular operator.

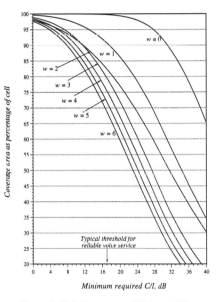

Figure 4 - Cell Coverage vs. Minimum C/I
($w =$ number of interferers, $N_0 = I_1 / 10$)

5.0 VOICE CHANNEL ASSIGNMENT ALGORITHMS

Voice users arrive at the base station at random intervals and request service through the mobile telephone switching office (MTSO). The MTSO assigns each new voice arrival to an idle channel (if one is available) according to some predetermined algorithm. If one ignores co-channel interference, the job of the hopping algorithm is to try to predict this assignment either statically or dynamically and hop to a channel that will be idle for the maximum time. Channel assignment algorithms fall into two categories: fixed channel assignment and dynamic channel assignment. We shall consider only fixed-channel assignment algorithms.

In a fixed channel assignment system, each base station is assigned m channels, numbered 1 to M. Consider two fixed channel assignment algorithms: random channel assignment and "least preferred."

A. Random Channel Assignment

When an arrival occurs, there are k busy channels, $k = 1, 2, 3, ..., m$, and m-k idle channels. If $k = m$, the system is busy and the arrival is blocked. If $k < m$, the MTSO assigns the arrival to one of the m-k idle channels randomly. In other words, each of the m-k channels is equally likely to get the arrival. Real cellular systems do not assign channels this way, but this algorithm approximates the condition where the hopping channel stream cannot predict the voice channel assignment. In other words, it is a worst case for the hopping channel stream since it has no useful information on where to hop to maximize dwell time.

The random assignment algorithm is also a useful model when the mobile data base station hops to the channel with the least interference. Assuming the assignment algorithm is independent of the co-channel interference level, channel assignment is random (i.e., equally likely) from the point-of-view of the mobile data base station.

It will be useful to know the probability that an arbitrary channel j is idle and thus available for data service. This probability can be derived rigorously by conditioning on the number of busy channels, or we can use the following intuitive argument: The rate at which arrivals are assigned to idle channels is simply $\lambda(1-P_m)$. The proportion of these arrivals that are assigned to a particular

channel, j, is just $\lambda(1-P_m)/m$ Thus, we can write

$$p = P(j \text{ is idle}) = 1 - \frac{E[k]}{m}$$

$$= 1 - \frac{\left(\frac{\lambda}{\mu}\right)(1 - P_m)}{m} \tag{6}$$

B. Least preferred

In this case, the MTSO designates one of the m channels the *least preferred* voice channel. This channel is used for voice traffic only when the remaining m-1 channels are busy. On the rare occasion when the least preferred channel is busy, the MTSO searches for the first idle channel and hands off the voice user to this new idle channel, thereby maximizing the idle time of the least preferred channel. The least preferred voice channel is the only channel used for data service. If the system cannot measure co-channel interference, this approach is optimal because it maximizes the dwell time of the channel stream and simplifies the channel hopping protocol to the on/off case. The simplest implementation of the least preferred voice channel is to use the same radio frequency channel at all co-channel cells, although different least-preferred radio frequency channels can be used at co-channel cells.

6.0 OPTIMAL HOPPING STRATEGIES

The optimal hopping strategy depends on whether the system can measure co-channel interference.

A. System Cannot Measure Interference

In this case, the optimal strategy is to designate a least preferred voice channel and use this channel exclusively for data service, when it is idle. To minimize co-channel interference to voice users, the same radio frequency channel can be designated least preferred voice channel at each co-channel cell. The voice user that is assigned to this channel would experience maximum co-channel interference since the channel would be used by either the data channel stream or another voice user at all times in all co-channel cells. However, the maximum co-channel interference would occur for an average of only $1/m\mu$ seconds, the mean time before another voice channel is idle.[2] Data service, on the other hand, would experience maximum co-channel interference on the forward channel (base to mobile) at all times.

B. System Can Measure Interference

The performance of the channel stream is limited by co-channel interference. Since the co-channel activity is beyond our control, we cannot reduce co channel interference in an absolute sense. We can, however, direct the system to hop to the idle channel with the least co-channel interference (or equivalently, the idle channel with the least number of active co-channels).

In the next section, we will need the probability mass function (pmf) for the number of busy co-channels for this hopping approach. To find the pmf, we assume that all seven co-channel cells are experiencing the same offered traffic and operate at the same transmit power levels. Consider the case of a single channel, j, $j = 1, 2, ..., m$, at one of the co channel cells. In Section 5, we found that an arbitrary channel is idle with probability p given by (6). Assuming that the busy/idle status of channel j is independent of its status in co-channel cells, the probability that l or fewer co-channels ($l = 1, 2, ..., n$) of channel j are busy is

$$F_r(l) = P(L \leq l) = \sum_{i=0}^{l} \binom{n}{i} p^i (1-p)^{n-i} \tag{7}$$

where $n = 6$ for an omnidirectional cell and $n = 2$ for a directional cell.

Let Q = the number of idle co-channels for the hopping channel. In other words, Q is the number

[2]An Erlang loss system with m busy servers and service rate μ acts like a single server with service rate $m\mu$. Therefore, the mean wait time for a service completion is simply $1/m\mu$. See [11, pp. 158].

of idle co-channels for the particular voice channel, j_{max}, with the maximum number of idle co-channels. Now, when there are k busy channels at our base station, there are $m-k$ idle channels and the probability that channel j_{max} (the hopping channel) has l idle co-channels or fewer, is the probability that all $m-k$ idle channels have less than or equal to l idle co-channels. Thus,

$$P(Q \leq l \mid k) = F_r(l)^{m-k}$$

where $F_r(l)$ is the cumulative distribution function for the number of idle co channels, L. To find the probability $P(Q \leq l)$, we calculate the sum,

$$F(l) = P(Q \leq l) = \sum_{k=0}^{m} F_r(l)^{m-k} P(k) \tag{8}$$

where $P(k)$ is the probability that k channels are busy, which is given by (2). The probability mass function for Q is found from the cdf, $F(l)$ by

$$p_Q(0) = F(0)$$
$$p_Q(l) = F(l) - F(l-1), l = 1, 2, 3, ..., n \tag{9}$$

7.0 CELL-WIDE PERFORMANCE OF CHANNEL STREAM

In this section, we calculate the average performance of the channel stream over the entire cell at a particular level of offered traffic. Performance is measured in two ways: coverage and throughput.

A. Coverage

In practice, the packet radio system sets some threshold for reliable service based on packet-error rate. For example, the CDPD specification recommends that a channel with 10% packet-error rate or higher be designated unreliable [4]. When this error rate is exceeded, service is suspended until the C/I and packet-error rate improve. A typical value of minimum C/I is 24.5 dB.

The probability that a randomly selected point in the cell will have adequate C/I is expressed as

$$P(U > u_{min}) = \sum_{w=0}^{n} P(U > u_{min} \mid w) P(W = w) \qquad (10)$$

where the conditional probability, $P(U > u_{min} \mid w)$, is given by (5) and the the probability that w co-channels are busy, $P(W=w)$, depends on how the channel stream is implemented. We shall consider three implementations: single "least preferred" channel, conventional random channel selection, or optimal channel selection (least co-channel interference).

In the case where the channel stream uses the "least preferred" voice channel, we assume that the same radio frequency channel is used at all seven co-channel cells. Since the forward channel stream transmits continuously, the number of busy co-channels is constant at $w = n$ and the coverage is given by (5) with $w = n$.

The conventional channel stream hops between idle channels ignoring co-channel interference. Therefore, the probability that w co-channels are busy is just

$$P(W = w) = p^{n-w}(1-p)^w \qquad (11)$$

where p is the probability that a co-channel is idle, given by (6).

The optimal implementation is to select the channel with the least co-channel interference. The probability that w co-channels are busy for this approach is given by $p_Q(n-w)$, using (9) from Section 6.

Equation (11) is plotted in Figure 5 for an omnidirectional cell for the three channel stream implementations. Note that coverage for the optimal approach is immune to traffic until the

offered traffic reaches a level of roughly 20 Erlangs. At this point, the coverage has dropped to 95%, but the blocking probability is still very low, $P_b = 1.6 \times 10^{-10}$. This observation supports Lee's conclusion that the cellular system is limited by co-channel interference more so than blocking [5]. The "least preferred" channel approach offers a poor coverage value of 45% at all levels of offered traffic.

Interestingly, the greatest performance gains for the optimal hopping algorithm occur at medium traffic levels. This result runs counter to the conventional wisdom that channel hopping is most helpful on congested systems.

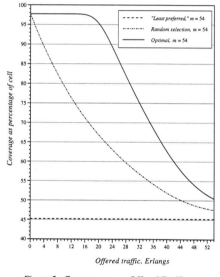

Figure 5 - Coverage versus Offered Traffic
(u_{min} = 24.5 dB, $N_0 = I_1 / 10$)

B. Throughput

To estimate throughput of the channel stream under the three implementations, we simulated a packetized transmission system with a Reed-Solomon code for error control. The parameters of this system are summarized in Table I. Note that the RS(63,47) code is capable of correcting up to $t = 8$ code symbol errors, but is used to correct only 7 errors to improve error detection performance.

The carrier-to-interference ratio is time-varying, but we assume that the ratio is constant for the duration of one bit. At the normalized (to symbol rate) doppler rate of $f_n = 0.0021$, this is a realistic assumption. The output of the simulation is an estimate of the probability of correct packet reception, P_{cd}, at a fixed ratio of mean energy per channel symbol to noise power spectral density, $\bar{\gamma}_s$. Assuming the receiver noise bandwidth is the channel symbol rate and the receiver has a noise figure of NF dB, the carrier-to-interference ratio, u, and $\bar{\gamma}_s$ are related by $u = \bar{\gamma}_s + \text{NF}$.

If repeated packets are separated sufficiently in time that they are uncorrelated, the mean data rate at a particular value of u in bits per second is given by [9, pp. 462]

$$\eta(u) = R_s P_{cd}(u) \tag{12}$$

where R_s is the channel symbol rate and we have assumed selective-repeat ARQ.

TABLE I - SIMULATED PACKET RADIO SYSTEM

Parameter	Value
Block (packet) size	378 bits
Symbol rate	19.2 ksps
Modulation	Binary FSK
Detection	Non-coherent, hard decision
Error Control	RS(63,47), GF(64), t=7
ARQ	Selective-repeat
Vehicle speed	50 kph
Channel model	Rayleigh
Carrier frequency	881 MHz

The mean data rate, $\eta(u)$, is plotted in Figure 6 for a noise figure of NF = 5 dB. Note that the 90% throughput threshold (10% packet error rate) occurs at $C/I = 24.5$ dB. We note that this threshold is 7.5 dB higher than the popular threshold for voice service (17 dB). Referring to Figure 4, we see the coverage difference between voice and data service

will be 28%. This is unfortunate because the user expects the data service availability to match the voice service availability. In practice, the difference will be even wider because cellular operators routinely lengthen the time-out on the supervisory audio tone (SAT) to extend the C/I voice threshold to 10 dB or lower.

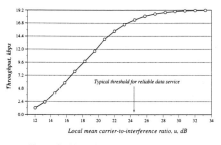

Figure 6 - Throughput in Rayleigh Fading

To determine the mean data rate over the entire cell, we compute

$$R_{avg} = E[\eta] = \int_0^\infty \eta(u) f(u) \, du \tag{13}$$

where $f(u)$ is the probability density function of the carrier-to-interference ratio.

Although $\eta(u)$ is a continuous function, our simulator furnishes only a discrete approximation for n_s values of carrier-to-interference ratio. Thus, we can approximate (13) by the following expression:

$$R_{avg} = \sum_{i=1}^{n_s+1} \eta[u(i)] \, [P(U > u(i-1)) - P(U > u(i))] \tag{14}$$

The probability, $P(U > u(i))$, is given by (10). The values at the boundaries are $u(0) = -\infty$ and $u(n_s + 1) = \infty$, so $P(U > u(0)) = 1$ and $P(U > u(n_s + 1)) = 0$.

Equation (14) is plotted in Figure 7 for the three hopping implementations where we have assumed $\eta(u) = 0$ for $u < 24.5$ dB.

One drawback of the optimal hopping approach is the greater number of hops per unit time. Since the mobile data base station is adapting to conditions in a total of seven cells rather than one, the number of hops will be roughly seven times that of a conventional base station. Most of these hops will take place during acceptable C/I levels, however, as opposed to the "forced hop" case when the base station hops at the onset of activity in the home cell. The impact of the greater number of hops depends on throughput los due to the "dead time" during the negotiation of a hop.

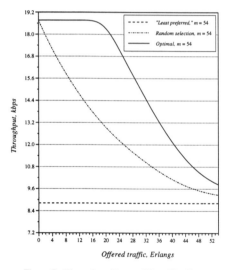

Figure 7 - Throughput Versus Offered Traffic
($m = 54$, $u_{min} = 24.5$ dB, $V = 50$ kph, $N_0 = I_1 /10$)

The cell coverage and throughput for omnidirectional and directional cells are summarized in Tables II and III for two levels of offered traffic. Note that by selecting the channel with the least co-channel interference, we can increase coverage in an omnidirectional cell by

47% and improve throughput by a factor of 2 over the "least preferred" channel implementation.

TABLE II - OMNIDIRECTIONAL CELL
($m = 54$, $R_s = 19.2$ ksps, SR ARQ, Min. $C/I = 24.5$ dB)

	$\lambda/\mu = 9.00$		$\lambda/\mu = 11.5$	
	Covg.	Tput.	Covg.	Tput.
L. Pref.	45.2%	8.82 kbps	45.2%	8.82 kbps
Ran. Sel.	64.8%	12.6 kbps	50.1%	9.77 kbps
Opt. System	92.6%	17.8 kbps	57.8%	11.2 kbps

TABLE III - DIRECTIONAL CELL
($m = 18$, $R_s = 19.2$ ksps, SR ARQ, Min. $C/I = 24.5$ dB)

	$\lambda/\mu = 9.00$		$\lambda/\mu = 11.5$	
	Covg.	Tput.	Covg.	Tput.
L. Pref.	66.3%	12.9 kbps	66.3%	12.9 kbps
Ran. Sel.	79.7%	15.4 kbps	75.8%	14.7 kbps
Opt. System	95.4%	18.3 kbps	88.9%	17.1 kbps

8.0 CONCLUSIONS

A cost-effective method to provide data service on the cellular radio network is to hop the data channel stream between idle radio channels. By doing so, one can achieve data service availability of $1 - P_b$ without increasing call blocking to voice users. A hopping channel stream also creates a new revenue-producing channel without the need for additional radio spectrum.

A hopping channel stream will cause additional average co-channel interference, but worst-case interference will be not be affected since the channel stream must hop off the system entirely when all channels are needed for voice use. A single forward (base to mobile) channel stream in a directional cell will increase average co-channel interference by 0.3 dB, resulting in 0.8% loss in geographical coverage.

Systems that cannot measure co-channel interference do not gain performance by channel hopping. Instead, these systems should use a single channel that is normally dedicated to data service and hop off this channel whenever it is needed for voice service. On the other hand, a system that can measure co-channel interference should select the idle channel with the least interference to maximize geographical coverage and throughput.

By choosing the channel with least interference in an omnidirectional cell, geographical coverage will increase by 47% and throughput will improve by a factor of 2 over the single channel case. In a directional cell, coverage will increase by 29% and throughput will improve by 42%.

Contrary to conventional wisdom, hopping offers the greatest capacity gains at medium traffic levels, not high traffic levels.

A single channel (vice hopping) approach to CDPD will result in high co-channel interference levels at mobile data receivers and poor geographical coverage. In fact, data service will be available in significantly fewer locations than voice service. Smart hopping strategies are needed to boost the availability of data service to the levels of the current voice system.

ACKNOWLEDGMENTS

Professor Mark Wickert of the University of Colorado at Colorado Springs developed the Rayleigh fading channel model used in the simulations. Professor Kevin Sowerby of the University of Auckland pointed out the effect of multiple interferers on the mean and variance of the carrier-to-interference ratio. Hermon Pon and Ken Picot of Northern Telecom provided helpful discussions.

APPENDIX - DERIVATION OF (5)

The mobile user receives interference from up to six co-channel base stations in addition to the desired signal from his or her home base station. All cells transmit equal power in equal bandwidths, serving identical cells of radius r_0. The mobile user's position, (r, θ), is uniformly distributed over the area of his or her home cell. We want the probability that the carrier-to-interference ratio, U, is greater than a threshold, u_{min}, given w co-channels are active. This probability can be written as

$$P(U > u_{min} \mid w) = 1 -$$

$$\frac{1}{\pi r_0^2} \int_0^{2\pi} \int_0^{r_0} P(U \leq u_{min} \mid w, r, \theta) r \, dr \, d\theta \qquad \text{(A-1)}$$

where we have assumed the cell is a circle of radius r_0. The random variable U is the difference of the desired signal power, C, and the total interference power, I, $U = C\text{-}I$. Both C and I are independent normal random variables, thus U is also normal [7]. The mean of U is the difference of the mean desired signal power, m_C, and the mean interference power, m_I, at position (r, θ) in the home cell,

$$m_U(s, \theta) = m_C - m_I \qquad \text{(A-2)}$$

Assuming C and I are independent, the variance of U is given by [7]

$$\sigma^2 = \sigma_C^2 + \sigma_I^2 \qquad \text{(A-3)}$$

where σ_C is the standard deviation of the desired signal power and σ_I is the standard deviation of the total interference power. The standard deviation of the desired signal power is typically between 6 and 12 dB for practical channels.

The mean of the desired signal as a function of position is given by

$$m_C(r, \theta) = k_t - \gamma \log_{10}(r) \qquad \text{(A-4)}$$

where $\gamma/10$ is the propagation exponent and k_t is a constant common to all base stations in the system.

Finding the mean and standard deviation of the interference power is a bit more involved because the interference power is the sum of w lognormal random variables. Let L_i be the lognormally-distributed power from a single interferer. Then

$$X_i = 10\log_{10} L_i$$

is the normally distributed power in dB from a single interferer with mean m_{x_i} and standard deviation σ_{x_i}. It will be convenient to work with natural logarithms. Define the normal random variable $Y_i = \ln L_i$ with mean m_{y_i} and standard deviation σ_{y_i}. The normal random variables X_i and Y_i are related by

$$Y_i = \kappa X_i, \quad m_{y_i} = \kappa m_{x_i}, \quad \sigma_{y_i} = \kappa \sigma_{x_i}$$

where $\kappa = 0.1\ln(10) = 0.23026$. The problem is to find the sum of w lognormal random variables, $L = L_1 + L_2 + ... + L_w$ or equivalently,

$$L = e^{Y_1} + e^{Y_2} + ... + e^{Y_w} \cong e^Z$$

where the interference $I = Z/\kappa$. Note that we are assuming that the sum of lognormals is also a lognormal. To find the mean and variance of Z (and in turn, I) we employ Wilkinson's method, described in [10, pp.165]. At this point, we still need an expression for the mean of a single interferer, m_{x_i}, in terms of r and θ. From Figure A-1 and the law of cosines, we can calculate the distance from the ith interferer to the mobile unit as

$$d_i(r,\theta) = \left(r^2 - 2d_0 r_0 \cos\left(\theta - i\frac{\pi}{3}\right)r + d_0^2 r_0^2 \right)^{1/2}$$

where d_0 is the normalized frequency reuse distance ($d_0 = 4.6$ for a 7 cell reuse pattern). The factor of $i\pi/3$ results from the 60° spacing of co-channel interferers (see Figure 3).

The mean power level from the ith interferer is given by

$$m_{x_i}(r, \theta) = k_t - \gamma \log_{10}[(d_i(r, \theta)] \tag{A-5}$$

By re-writing $P(U \leq u_{min}|w, r, \theta)$ in terms of the unit normal distribution and making a change of variables, $s = r/r_0$, we get (5) of Section 4.

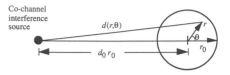

Figure A-1 - Co-Channel Interference Geometry

REFERENCES

[1]   W.C. Jakes, ed., Microwave Mobile Communications, IEEE Press Reissue, 1994.

[2] V. H. MacDonald, "The cellular concept," The Bell System Technical Journal, January 1979.

[3] W.C.Y. Lee, Mobile Cellular Telecommunications Systems, New York: McGraw-Hill, 1989.

[4] Cellular Digital Packet Data System Specification, Release 1.1, January 1995.

[5] W.C.Y. Lee, "Data Transmission via Cellular Systems," 43rd IEEE Vehicular Technology Conference, May 18-20, 1993, Secaucus, New Jersey.

[6]   N. Abramson, "The throughput of packet broadcasting channels," IEEE Trans. Commun., pp. 117-128, Jan 1977.

[7] S. M. Ross, A First Course in Probability, New York: MacMillan, 1984.

[8] D.R. Cox and W.L. Smith, Queues, London: Chapman and Hall, 1961.

[9] S. Lin and D. J. Costello, Error Control Coding: Fundamentals and Applications, Prentice Hall, 1983.

[10] A.A. Abu-Dayya and N.C. Beaulieu, "Outage probabilities in the presence of correlated lognormal interferers," IEEE Transactions on Vehicular Technology, pp. 164-173, February 1994.

[11] G.F. Newell, Applications of Queueing Theory, 2d Ed., London: Chapman and Hall, 1982.

A CSMA/CA Protocol for
Wireless Desktop Communications

Jim Lansford
Momentum Microsystems
2864 So. Circle Drive, Suite 401
Colorado Springs, CO 80906
+1 719 540 8338 • +1 719 540-8361 fax
support@mmicro.com

Abstract

There is an aggressive push among computer companies to consolidate and enhance peripheral communications. One group, the ACCESS.bus Industry Group, has developed a serial bus architecture that offers significant technical and cost advantages over the existing peripheral ports. This industry group is now taking the next step: wireless desktop communications. The paper will discuss issues in serial bus protocol design, and shows both network and signal level simulations of the proposed standard for CSMA/CA wireless desktop communications.

Serial bus technology is expected to evolve rapidly over the next few months; the paper will conclude with a summary of higher performance buses that are on the horizon.

1.0 Background

As portable computing devices shrink in size, the current hodgepodge of connectors required to attach mice, printers, scanners, trackballs, keyboards, keypads, peer-to-peer interfaces, etc. are increasingly difficult to justify because of their cost, size, reliability, and limited connectivity. An alternative has emerged, and promises to be an open, industry standard desktop bus. ACCESS.bus (A.b) allows up to 125 devices to be addressed over a single 100kb/s serial bus, with higher speeds as an option; A.b was developed as a low-cost desktop bus that would replace the keypad, parallel, serial, mouse, and keyboard connectors with a single connector, while greatly expanding the number of desktop devices that can be attached to a system This bus can be though of as the backbone of what can be called a desktop area network, or piconet. The bus standard is fairly mature, and offers a number of unique features, including "hot plug-in", which provides for devices to be plugged and unplugged while the host system is active. Access.bus represents a low cost, lower performance network than 802.x (but is not meant to compete with it) and is an ideal solution for the problems of desktop connectivity - printers, input devices, and short range peer-to-peer transfers.

The general topology of a Wired ACCESS.bus system is given in Figure 1, when details of the wired protocol can be obtained from the ACCESS.bus Industry Group [1].

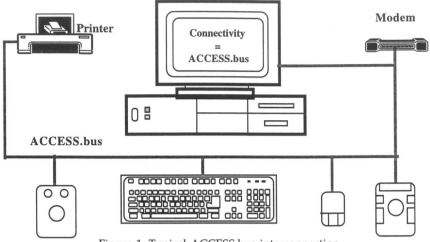

Figure 1: Typical ACCESS.bus interconnection

Key to widespread acceptance of A.b for the subnotebook and PDA markets is a wireless implementation; the Access Bus Industry Group is developing a CSMA protocol for RF wireless A.b as well as an infrared protocol for point-to-point use. A proposed RF solution uses conventional RF hardware and modulation techniques to implement a very low cost, robust, multiple access protocol. The protocol uses conventional packet radio techniques based on a tightly coupled embedded microcontroller. The optional error correction scheme is based on a block Hamming code; a CRC over the block is the default that assures a robust detection of data corruption. An ALOHA scheme of handshaking forces the receiver to acknowledge receipt of a correct block; the transmitter retries until an acknowledgment is received, or passes an error message back to the host if the transaction could not be successfully completed.

2.0 Wireless ACCESS.bus - overview

A draft specification of the proposed protocol is currently under review by members of the ACCESS.bus Industry Group; the packet format is given in Figure 2. The initial four bits are a 1010 pattern that is used to derive bit synchronization; the following four bits are packet options that allow enhancements to the basic protocol. The first bit, when set, enables an optional error correction algorithm that is implemented in addition to the basic cyclic redundancy check (CRC) that is the normal error detection scheme. The next bit enables an optional data compression algorithm, which is not fully defined; it could be as simple as a run length encoding algorithm, or a more elaborate Huffman-style compression scheme. The third aand fourth bits are reserved for future uses such as encryption, or to enable a compatibility mode with a future protocol.

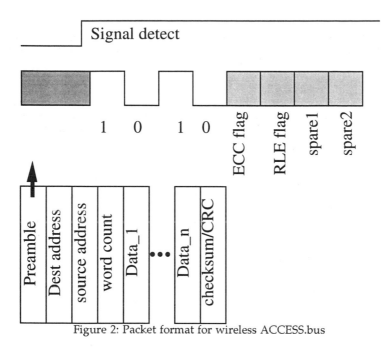

Figure 2: Packet format for wireless ACCESS.bus

3.0 Overview of physical layer

The ACCESS.bus wireless systems will employ low cost, conventional RF and infrared technology to provide wireless links to peer and peripheral devices at speeds up to 150kb/s. The basic protocol is CSMA/CA with unslotted handshaking (ALOHA), where only ACK messages (not ACK/NAK) are sent as a response. The ACCESS.bus wireless has been initially designed with an RF physical layer, but with IR options.

3.1 RF physical layer

The RF physical layer will employ a portion of the RF spectrum set aside in the United States for Industrial, Scientific, and Medical (ISM) users. We have selected the ISM band that allows a highly cost effective solution. The ACCESS.bus will operate under §15 of the FCC rules, which does not require a license; only type approval is necessary. One clarification: there are provisions for unlicensed and licensed devices in this band that allow higher power transmission but require much more expensive spread spectrum technology. The ACCESS.bus wireless does not operate under these rules; the §15.249 rules allow for low power (<50mV/meter at 3 meters) operation virtually unrestricted.

3.1.1 Frequency

The ACCESS.bus wireless system will use the 902-928MHz ISM band, which allows unlicensed users to transmit at field strengths of up to 50mV/meter at a distance of three meters from the antenna. Frequency availability in other countries is under study.

3.1.2 Bandwidth and Data Rate

Based on usage by existing devices in the 902-928MHz band, the ACCESS.bus wireless will operate at approximately 906MHz, and will occupy a bandwidth of approximately 280kHz centered at that frequency. The field strength outside of this 902-928MHz band must be no more than $500\mu V$/meter at distances greater than 30 meters from the antenna (§15.209). The data rate will be 150kb/s.

3.1.3 Multiple Access plan

The proposed ACCESS.bus wireless system uses a carrier sense multiple access with collision avoidance (CSMA/CA) protocol to allow multiple devices to share the band. The typical scenario would be for a device to listen for a period of no traffic, then begin its transmission. If another device began transmitting simultaneously, then the intended receiver would receive a garbled message, since all users share the same frequency. Each successful transmission is followed by an acknowledgment at an unspecified later time over the same channel; this is called ARQ or ALOHA. This scheme was chosen because it maximizes data bandwidth for low duty cycle or low contention transmissions, which will be assumed to comprise the majority of transactions on the network; the goal has been to provide maximum performance consistent with low cost. Momentum is in the process of developing detailed signal models to insure that this CSMA/CA scheme will degrade gracefully as message traffic increases.

3.1.4 Modulation plan

The modulation will be FSK, with the mark/space deviations to be resolved; the preliminary deviation specification is 100kHz. Thus, the mark frequency will be 905.95MHz and the space frequency will be 906.05MHz.

3.1.5 Power level

Assuming a unity gain (0dB) antenna, the field strength allowed by the FCC (50millivolts/meter) translates into a received power level of -20 dBm at 3 meters, which implies an effective radiated power of -1.25dBm ($700\mu W$).

3.2 Implementation issues
3.2.1 Antenna

An antenna is application specific, but could easily be implemented as a

PC board trace; other antenna possibilities are a loop (for a mouse) or even an external whip.

3.2.2 Receiver architectures

For FSK, there are a number of off-the-shelf receiver chips which implement an IF section and discriminator or PLL in a single chip with a minimum of external passive devices. A limiter should be used to more sharply define the usable RF field and to reject interference within the operating radius. The FSK receiver architecture is implementation specific and will not be delineated in this document.

3.2.3 Range

Range will be 30 meters minimum, based on propagation models given by Motley and Keenan [2]; propagation beyond that range has not been accurately predicted, but a 50 meter range is considered typical for a normal office environment. The antenna pattern may be hemispherical or torodial, depending on the type. Polarization is not expected to be critical given the environment, but will be nominally horizontal.

3.2.4 Power management strategies

Since single channel RF wireless devices are half duplex (either transmitting or receiving but never both), some power management in the RF circuitry is inherent. When the transmit/receive (T/R) switch is in receive, the transmitter will be turned off, and vice versa. The μC will also have power down modes, although these are not yet defined, since the choice of μC will be left up to individual vendors. For small battery powered devices such as a mouse, a deep sleep mode may be necessary, where the device shuts itself off unless interrupted (movement of a mouse, for example). This may pose a problem, since a host could flag a message as undeliverable, even though the device is within range. (See MAC spec)

4.0 Media access control (MAC) layer

The MAC layer defines the structure of the message packet and how the integrity of the information within the packets is assured. The ACCESS.bus wireless protocol is similar to the wired ACCESS.bus format, with the addition of a synchronization preamble and a more robust error detection and correction scheme. Error detection is implemented using a CRC; error correction can be optionally employed to make the system more robust. The ACCESS.bus wireless also supports peer to peer transfers.

4.1 Message format (device transfers)

While there can be a lot of similarity in the packet structure between a wired bus and a wireless system, there are some significant differences:

a) Since there is no separate clock line in a wireless link, bit synchronization or clock recovery must be used to reconstruct any clock signals at the receiver.

b) A wireless link is much more likely to incur errors during transmission, so an error protection method which is more robust than a checksum is required in many instances.

c) A wireless link has much greater latency than a wired link, so byte handshaking is impractical.

With these constraints in mind, the wireless packet will have three components: a preamble which will allow clock recovery and bit synchronization and provide for some wireless options, the information bundle, consisting of a destination address, source address, word count, data, and a CRC trailer. Each of these portions will be described in the following sections.

4.1.1 Preamble

The wireless packet begins with a preamble sequence that is used to recover the clock in the receiver; this clock recovery can be though of as similar to a phase-locked loop, where the approximate timing is known, but the "phase" (the exact time the center of the bit occurs) is not known. It is proposed to define an 8-bit preamble, where the first 4 (four) bits are the sequence [1 0 1 0]; this sequence should be adequate to recover the clock timing in all cases except those where a collision has occurred.

4.1.2 Optional fields

In an effort to provide for future enhancements, we propose that the four additional bits in the header be used to select hardware/firmware specific options in the packet. Thus, the entire preamble and suggested assignments would be as follows:

$$[1\ 0\ 1\ 0\ a\ b\ c\ d]$$

a = error correction coding (ECC)
b = run length encoding (RLE)
c = spare (USB, encryption, etc.)
d = spare

4.1.3 Encapsulated data packet

After the preamble, the packet will consist of a synchronous data burst (no start/stop bits) of destination address, source address, word count, data bytes, end of packet CRC

4.1.4 End of transmission - CRC

The end of a transmission will be a marked by a CRC block; if chosen properly, the CRC should allow detection of multiple bit errors, including bursts. The proposed polynomial is $X^{16}+X^{15}+X^2+1$; implementation of this algorithm will be in the MAC μC. This polynomial guarantees detection of error bursts of up to 16 bits in length.

4.2 Wireless "plug in" and host ownership

The process of plugging devices into the wireless "bus" involves what will be called the "discovery" process. Like HDLC, host (primary) devices will issue a "sniffing" packet periodically (TBD - probably every few seconds) that will invite devices or other hosts to "plug in" to the wireless "bus". Any device hearing the discovery packet would check to see if that host was known or unknown; if known, the packet is ignored. If the discovery packet is from an unknown host, the device or host sends an acknowledgment, including its 32-bit ID code. The sniffer would then send a packet asking for a capabilities, to which the new device would respond with a capabilities string, including a list of current device address assignments. The sniffer would respond to a new host device with its own list of device assignments, which would allow them to share devices within their domains.

4.3 Peer-to-peer protocol

The recommended implementation is definition of a new device class, which should be called "Peer" or "Host". Definition of the attribute string associated with a device type of "Peer" will need to be developed in conjunction with existing desktop bus software developers, but would likely include host type, IP address (if defined), operating system type and version (DOS, DOS/WIN, MAC, UNIX, etc), ACCESS.bus firmware revision, ECC capability, RLE capability, encryption capability, USB, etc. Other issues are that the files in peer to peer transfers can be larger, and that packets in a file transfer should be sequentially numbered so that the recipient can request retransmission of a particular packet (like XMODEM). Of critical importance is a way to prevent a peer to peer transfer from blocking other devices; software developers for ACCESS.bus have been addressing these issues of "device throttling" and wireless ACCESS.bus should to be able to adapt these methods.

A fundamental issue is negotiation of the communication between peers; ACCESS.bus is a single host protocol, where the host is defined to have an address of 50h. In order to allow peer-to-peer transfers, a possible solution is to allow all transactions between any peripheral and its host to occur between its device address and 50h. Thus, a peripheral will always send messages to 50h, and the receiving host must decide whether to process the packet further. For transfers between hosts, the hosts could decide to communicate on alternate addresses; this negotiation would take place at initialization time, after capabilities strings are exchanged. This would allow hosts to communicate without interfering with conventional ACCESS.bus transactions. The choice of addresses for these peer-to-peer transactions is arbitrary, and would be incorporated into an enhanced bus manager.

4.4 Undeliverable message disposition

The disposition of undeliverable messages must depend on the type of device the message is addressed to. For example, if a peer to peer transfer is taking place and suddenly the sender fails to receive acknowledgments from the recipient, then the link has been disrupted and an error message should be displayed after several attempts have been made. On the other hand, if a host issues a command to a device that has gone into a deep sleep mode, it must either presume the device has gone to sleep, been turned off, or moved out of range.

2.5 Arbitration or collision detection protocol

A CSMA/CA wireless protocol is a "listen before talking" scheme - RF devices cannot emulate the wired-AND or twisted pair schemes a wired bus might use for arbitration. A device (or peer) wishing to send a packet would wait until the end of a transmission, then if it hears no activity, would start transmission of a packet. If, for some reason, another device began transmission during this packet, the data could be corrupted, which would be indicated by an erroneous CRC field (such a situation can easily occur when there are "hidden" nodes - when two devices can communicate with the host, but cannot see detect each other). Since this is an ACK system (not an ACK/NAK), the packet would not be acknowledged, indicating to the sender that the packet was not successfully received. When this occurs, the sender should wait a period of time before attempting retransmission (called the backoff); for ACCESS.bus wireless, this backoff time will be proportional to the device address (1-125), and the initial proposal is ADDRESS*2.5mS. This is consistent with wired ACCESS.bus, because lower addresses take priority over higher ones; the specific choice of 2.5mS arose because hidden nodes force a choice of backoff that will allow a 32-byte packet to be transmitted from consecutive addresses without collision, creating an ersatz slotting scheme. The backoff algorithm can be used in conjunction with bandwidth management features to control devices and peers that are on the air an excessive amount.

Summary

This paper has presented an overview of a new wireless protocol that is based on an entirely different paradigm than most wireless communications devices for computers - it is designed for the low end, peer to peripheral or light duty peer to peer communications using a serial bus protocol. As such, it is far cheaper to implement than traditional wireless systems that are based on an ethernet protocol, and offer the possibility to design RF wireless communications into devices that could not have previously tolerated the cost.

References

[1] "ACCESS.bus Rev 2.2 Specification", available from ACCESS.bus Industry Group, 370 Altair Way, Suite 215, Sunnyvale, California 94086
[2] See *The Mobile Radio Propagation Channel* , by David Parsons, Wiley, 1992, pp. 194-197.

CDMA Forward Link Capacity in a Flat Fading Channel

Louay M.A. Jalloul and Kamyar Rohani

Motorola Inc., Fort Worth Research

5555 N. Beach St., Ft. Worth, TX 76137

Abstract—*The performance of the forward link for an IS–95 based CDMA system in case of slowly moving mobiles in a flat Rayleigh fading channel is evaluated. It is shown that soft handoff improves the system capacity at the expense of an increase in the hardware requirements at the cell site. The analysis shows the interaction between the design parameters and their effect on system performance. Outage, blocking and coverage probabilities are defined and used to characterize system performance.*

1. INTRODUCTION

In the past few years we have seen a rapid growth in the demand for wireless communication services, thus creating the need for systems that offer higher capacity than the existing analog system. Code Division Multiple Access (CDMA) is emerging as a viable solution to increase the capacity and quality of service for cellular mobile radio systems [1,2]. IS–95 is the wideband digital cellular standard based on CDMA [3]. Cellualr radio system designs are typically characterized by cells with large coverage areas, servicing mobiles that travel at high speeds. A modification of the IS–95 CDMA standard is also being considered for a new market in wireless communications, namely personal communication systems (PCS). PCS systems differ from cellular systems in two main attributes. First, PCS service areas are densely populated and subcribers typically move at relatively low speeds, thus smaller cells (also known as microcells) are used. Second, propagation measurements in microcells have shown that delay spread is very small [4]. Hence, the objective of this paper is to examine system performance for slow mobile speeds and flat fading conditions.

While most analyses of CDMA have focussed on the reverse link (mobile–to–base) [5 9], less attention has been paid to the forward link (base–to–mobile). This is attributed to the belief that CDMA systems are reverse link limited [1,2,5,6]. In the CDMA digital cellular standard (IS–95), the reverse link uses a *64*–ary orthogonal modulation, rate *1/3* constraint length *9* convolutional code, combination of open loop and closed loop power control, 2–fold (spatial) antenna diversity and a *4*–finger RAKE receiver. On the other hand, the forward link uses pilot–assisted coherent binary phase shift keying (BPSK), rate *1/2* constraint length *9* convolutional code, no antenna diversity and only a *3*–finger RAKE receiver. Although, the reverse link receiver has to perform noncoherent demodulation, it was shown that an E_b/N_o of *5.8* dB is needed to maintain a frame erasure rate (FER) of *1* percent in a flat Rayleigh fading channel at *100*Kmph [6]. Note that receiver performance on the reverse link improves for low Doppler frequencies [6]. This is due to fast power control (updating at *800*Hz) which is fast enough to partially compensate for the effects of channel fading.

Unlike the reverse link, the performance on the forward link degrades for slow mobile speeds. At low Doppler frequencies fade durations may be too long, thus reducing the benefits of interleaving and convolutional coding. Furthermore, receiver performance becomes even worse when the channel is flat Rayleigh fading, i.e. there is no multipath (RAKE) diversity benefit. Therefore, if all users in the CDMA system move slowly (below *5* mph), one would expect that system capacity should drop substantially. However, due to soft handoff system capacity can be restored even under the harsh condition of flat fading and slow mobile speeds.

On the forward link power control takes the form of power allocation [1]. Each cell site transmits the minimum amount of power such that the user meets a target FER. The transmitted power is inversely proportional to the signal–to–interference ratio seen at the mobile receiver, hence, reducing the interference caused to other users in other cells and thus increasing the overall system capacity. This method of power allocation results in all users having the same average FER.

A simulation of the CDMA forward link was carried out in [10], and the sensitivity of the system capacity to some of the design parameters was shown. The forward and reverse link capacities have been evaluated in [11] for multipath channels and for systems with transmit diversity used to combat flat Rayleigh fading. The studies in [10] and [11] both assumed a fixed E_b/N_o for the simulation of the forward link. However, the Eb/No on the forward link is highly a function of the soft handoff condition. The development of a CDMA system planning tool is described in [12], however, this study doesn't provide capacity evaluations.

Our aim in this paper is not to develop a system planning tool that models every detailed aspect of IS–95. Instead, we develop a framework for analyzing system performance. This framework allows us to identify key parameters that have a significant effect on system capacity. Also, insight is gained into the interaction between these design parameters. A semi–analytic model for analyzing the CDMA forward link performance in a flat fading channel is developed. Since the model assumes that the mobile is moving slowly, then fading is constant over a frame. Thus, allowing a quasi–static receiver to be integrated with the system simulation.

This paper is organized as follows. Section 2 describes the system model and assumptions used in the paper. An analysis of the effect of soft handoff on power allocation is shown in Section 3. The system performance measures are defined in Section 4 which also contains the discussion of the simulation results. Finally, Section 5 has some concluding remarks.

2. SYSTEM MODEL

The total received power by a mobile from the jth cell site is given by,

$$P_j = d_j^{-\mu}\ 10^{\xi_j/10}$$

where d_j is the distance from the mobile to the jth cell site, μ is the path loss exponent and ξ_j is a Gaussian distributed random variable with zero mean and standard deviation σ , representing shadow fading. Typically, σ takes values between 6.5 and 8. As in [13], assume that ξ_j may be expressed as the sum of a component in the near field ζ, common to all cell sites, and a component in the far field which is independent from one cell site to the next ζ_j. Both components are assumed to be Gaussian distributed random variables with zero mean and standard deviation σ independent from each other, so that

$$\xi_j = a\zeta + b\zeta_j$$

where

$$a^2 + b^2 = 1.$$

Let K denote the number of cell sites in the system. Assume that $K=19$, i.e. the system consists of a center cell plus 2 rings of interfering cells around it. Since the interference contributed by cells beyond the second ring is negligible, they are not considered in this analysis. Users are assumed to be uniformly distributed over the cell area which is modeled as a hexagon. Omnidirectional antennas are used at the cell sites and all cells transmit the same amount of total power. Let $(1 - \beta)$ be the fraction of cell site transmitted power that is dedicated to Pilot, Paging and Sync channels. Thus, β denotes the fraction of the total transmitted cell site power dedicated to all traffic channels.

The user arrival process is assumed to be a homogenous Poisson process with mean denoted by N_s, so that the offered load excludes the increase caused by soft handoff. The increase in the offered load (overhead) due to soft handoff is discussed later in the paper.

The channel is assumed to be flat Rayleigh fading and all users are moving slowly, e.g. at *5 mph* or lower. Thus, the Doppler frequency is less than *10 Hz* at a carrier frequency of *900MHz*. Since, the frame duration in IS–95 is *20 ms*, i.e. *50* frames per second, then the variations in the received signal envelope due to short term fading are at least *5* times lower than the frame rate. Therefore, at slow mobile speeds, fading can be assumed constant over a frame. Thus, the FER is given by

$$FER = \int_0^\infty P_f(\gamma) \; p(\gamma) \; d\gamma \qquad (1)$$

where $P_f(\gamma)$ is the conditional *FER* derived for a static AWGN channel, γ is the instantaneous bit energy–to–interference plus noise ratio and $p(\gamma)$ is the p.d.f. of γ. Note that γ and $p(\gamma)$ depend on the number of RAKE fingers locked at the mobile receiver.

Since mobiles are moving slowly, variations in path loss and lognomal shadowing are small and hence are neglected, only variations due to Rayleigh fading are considered. Also, the effect of thermal noise is neglected in the analysis since it is assumed that the power transmitted by the cell site is much higher than the thermal noise floor. There are many implementation and system losses not accounted for in the our analysis. Thus, the results presented here provide an upper bound on the actual system capacity in a flat slow fading channel. Table 1 lists the values of the simulation input parameters.

Table 1. Simulation input parameters.

Information bit rate (Kbps)	9.6
Path Loss Exponent	4
Lognormal Standard Deviation (dB)	8
Site–to–Site Correlation (%)	50
Pilot, Paging & Sync channels (% of total power)	30
Voice activity factor (%)	40
Target *FER* (%)	1
Processing gain (dB)	21
Static E_b/N_o at *1% FER* (dB)	3
Number of cell sites	19
Antenna	Omnidirectional

3. FORWARD LINK ANALYSIS

Before presenting the simulation results, a simple analysis is given that provides some insight into the effect of soft handoff on system capacity.

2.1 Effect of Soft Handoff on Power Allocation

Consider a system with only two cells and a mobile that is located between the two cells such that the average power received from each cell is the same. Let ϕ denote the fraction of power allocated by cell-1 to the mobile's traffic channel. In this case γ may be written as

$$\gamma = \frac{W}{R}\beta\phi\left(\frac{r_1^2}{r_2^2}\right) \tag{2}$$

where W/R is the processing gain, r_1 and r_2 denote the envelopes of the received signals from cell sites 1 and 2, respectively. Assuming that r_1 and r_2 are independent Rayleigh distributed with an rms value of unity, then it can be easily shown that the average γ is given by

$$\bar{\gamma} = \frac{W}{R}\beta\phi \tag{3}$$

In order to achieve a FER of 1%, $\bar{\gamma}$ is approximately 23dB[1]. Substituting the value of $\bar{\gamma}$ in Eq. (3) and solving for ϕ, we have

$$\phi = \frac{10^{23/10}}{0.7 \times 128} = 2.2 \tag{4}$$

Eq. (4) shows that the power allocated to a single user exceeds 1 and hence the required $\bar{\gamma}$ cannot be met.

Next, consider the case when the mobile is in soft handoff with two cells. In this case each RAKE finger is locked to a different cell site, thus providing the diversity effect. Since the forward link is coherent pilot assisted modulation, using maximal ratio combing, γ is given by

$$\gamma = \frac{W}{R}\beta\phi\left(\frac{r_1^2}{r_2^2} + \frac{r_2^2}{r_1^2}\right) \tag{5}$$

Note that the terms within the parentheses in Eq. (5) are the reciprocal of each other and this quantity is minimum when $r_1 = r_2$. Therefore, γ value can be lower bounded by

$$\gamma > \frac{W}{R}\beta\phi \times 2 \tag{6}$$

For a coherent BPSK modem with rate 1/2 constraint length 9 convolutional code and soft decision decoding, approximately 3 dB information E_b/N_o is needed to achieve a FER of 1% in a static AWGN channel. Hence, from Eq. (6), ϕ can be calculated as

$$\phi < \frac{10^{3/10}}{2 \times 0.7 \times 128} \approx 0.01 \tag{7}$$

Similarly, for the case of three way soft handoff, γ is given by

$$\gamma = \frac{W}{R}\beta\phi\left(\frac{r_1^2}{r_2^2 + r_3^2} + \frac{r_2^2}{r_1^2 + r_3^2} + \frac{r_3^2}{r_1^2 + r_2^2}\right) \tag{8}$$

which may be lower bounded by

---

1. This figure comes from the fact that there is a 20dB fade 1% of the time and a 3dB margin is needed at 1% FER.

$$\gamma > \frac{W}{R}\beta\phi \times 1.5 \qquad (9)$$

when $r_1 = r_2 = r_3$. Thus, the required power is

$$\phi < \frac{10^{3/10}}{1.5 \times 0.7 \times 128} = 0.015 \qquad (10)$$

Note that the power allocated to the mobile has dropped by a factor of about 100 due to soft handoff. However, the interference due to other cells has not been included in this analysis. This issue is discussed in the next section.

2.2 Soft Handoff

In this section the effect of the interference from the other cells on the power allocation is quantified. Let I_o denote the total interference power measured at the mobile,

$$I_o - \sum_{j=1}^{K} \rho_J \, P_J \qquad (11)$$

where ρ_j is a random variable that depends on the number of active mobiles connected to the jth cell site. Without loss of generality, assume that

$$P_1 > P_2 > P_3 > \ldots > P_K > 0 \qquad (12)$$

For a given mobile location, let M be the number of cell sites the mobile is in soft handoff with, i.e.

$$M = \begin{cases} \max(m), & \text{such that } \beta_{pilot}\dfrac{P_m}{I_o} > T_{ADD}, & \text{for } m = 1, 2, 3 \\ 0, & \text{otherwise} \end{cases} \qquad (13)$$

where T_{ADD} is a threshold that the Pilot strength from the cell site must exceed in order to be included in soft handoff[2] and β_{pilot} is the fraction of the total cell site transmitted power dedicated to the Pilot channel (assumed to be 20%). Note that the value of β_{pilot} used here makes the probability that M equals 0 almost negligible. Furthermore, let I_{oc} denote the interference power seen at the mobile caused by all cells excluding the ones in soft handoff, i.e.

$$I_{oc} = \sum_{j=M+1}^{K} \rho_j \, P_j \qquad (14)$$

Define q as the relative interference power seen at the mobile,

$$q \overset{\Delta}{=} \frac{I_{oc}}{\sum_{m=1}^{M} \rho_j \, P_m} \qquad (15)$$

In order to simplify the analysis we assume that $\rho_j = \rho$, $j = 1, 2, 3, \ldots, K$. Furthermore, we assume that

$$\rho = (1 - \beta) + \beta \, V \qquad (16)$$

The first term in Eq. (16) represents the fraction of the total power received due to Pilot, Paging and Sync channels, which is a fixed quantity, and V takes on two possible values

$$V = \begin{cases} 0.4 & \text{if the other cells are at mean power} \\ 1 & \text{if the other cells are at full power} \end{cases} \qquad (17)$$

---

2. This is not the conventional use of T_{ADD}. See [15] for the use of T_{ADD} in IS–95 pilot set management.

In order to examine the effect of the other cell interference on the power allocation, consider the two cell geometry of section 2.1. Let a denote the ratio of the average powers received from cell–1 to cell–2. Then, using the definition of q, γ is given by[3]

$$\gamma = \frac{W}{R}\beta\phi\left(\frac{r_1^2}{a\ r_2^2 + (1 + a)\ q} + \frac{a\ r_2^2}{r_1^2 + (1 + a)\ q}\right) \qquad (18)$$

Eq. (18) clearly shows that the power allocation ϕ increases as the relative interference increases. The CDF of q using the parameters in Table 1 for $V=0.4$ is plotted in Fig. 1. The statistics of q are given for three cases: $M=1$ (no handoff), $M=2$ (2–way soft handoff) and $M=3$ (3–way soft handoff). Fig. 1 shows that the relative interference is small when the mobile is not in soft handoff. In this case the mobile location is typically close to the cell site and thus its power allocation is small. As the distance between the mobile and the cell site increases the relative interference increases. However, the mobile typically enters soft handoff, which provides a diversity benefit and hence the power allocation is also small.

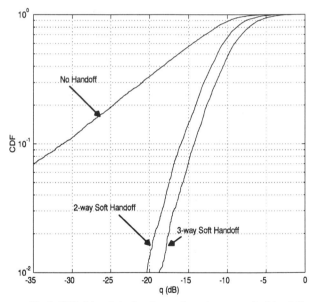

Fig. 1. CDF of the relative interference for various ways of soft handoff.

Table 2 shows the effect of increasing q on the power allocation. For each q value, ϕ was varied (using the quasi–static method) until a *FER* of *1%* was achieved. Clearly, power allocated to the mobile must increase as the other cell interference increases. Note that due to soft handoff, even with high values of q, the allocated power is less than 0.13. Table 3 shows the impact of imbalance on power allocation. As the imbalance increases, the diversity benefit due to the second RAKE finger is reduced.

---

3. Note that the other cell interference does not include variations due to Rayleigh fading, since it is the sum of many interferers modeled as Gaussian noise.

Table 2. Power allocation as a function of q for $\alpha=0$ dB.

q (dB)	ϕ
$-\infty$	0.0064
-12	0.012
-6	0.035
-3	0.065
0	0.13

Table 3. Power allocation as a function of imbalance α for $q=-3$dB.

α (dB)	ϕ
0	0.065
-3	0.068
-6	0.081
-9	0.097

4. SIMULATION MODEL AND RESULTS

A method described in [1] computes the forward link capacity for the case of no soft handoff with a fixed a E_b/N_o for all users in the system. Unlike this method, our scheme integrates both the mobile receiver characteristics and the system aspects of the CDMA forward link. Thus, it allows for multi-way soft handoff with variable E_b/N_o to meet a target *FER*.

4.1 Outage, Blocking, Coverage and Capacity

The instantaneous bit energy–to–interference ratio is given by

$$\gamma = \frac{W}{R}\beta\phi \sum_{m-1}^{M} \frac{P_m\ r_m^2}{\sum_{\substack{j=1 \\ j \neq m}}^{K} \rho\ P_j\ r_j^2} \tag{19}$$

where ϕ is the power allocated to the mobile such that the target *FER* is met (set to 1% for the simulation results). However, the system imposes upper and lower constraints on the power allocation, thus the *FER* of each user varies according to its location. The statistics of the *FER* are reported in the next section. Note that due to orthogonal Walsh covers on the forward link there is no interference among users' signals in the same cell.

Outage probability is defined as [1]

$$P_{out} = Prob\left[\sum_{i=1}^{N_h} \phi_i \nu_i > 1\right] \tag{20}$$

where ϕ_i and ν_i are the power allocated and the voice activity factor of the ith mobile, respectively. ν_i is modeled as a Bernoulli random variable taking values of 0 and 1 with probabilities of 0.6 and 0.4, respectively. N_h denotes the number of users served by a cell site which includes the offered load plus the overhead due to soft handoff. The IS–95 CDMA standard has only 64 Walsh covers on the forward link. Thus, the maximum allowable value of N_h plus the number of the overhead channels (Pilot, Paging and Sync) should not exceed 64. Allowing at least 4 channels for overhead, we limit the value of N_h to 60. Let N denote the offered load, which is Poisson distributed with mean N_s, then

$$N_h = \begin{cases} \sum_{i=1}^{N} M_i , & \text{if } \sum_{i=1}^{N} M_i < 60 \\ 60 , & \text{otherwise} \end{cases} \qquad (21)$$

where M_i denotes the number of cell sites the ith user is in soft handoff with. Thus, the blocking probability is defined as

$$P_{block} = Prob(N_h > 60) \qquad (22)$$

Coverage probability is defined as the fraction of the area for which the system average *FER* is less than a given threshold (typically 1% is used).

Capacity, measured in Erlangs, is defined as the maximum possible offered load for a given outage, blocking and coverage probabilities. In this paper, an outage of 1%, blocking of 1% and coverage of 90% are used to compute the capacity.

4.2 Results

Fig. 2 shows a plot of the outage probability versus the offered load for two cases: (1) the other cells are at mean power, i.e. *V=0.4* and (2) the other cells are at full power, i.e. *V=1*. Fig. 3 shows a plot of the blocking probability when the other cells are at mean power. As expected outage and blocking increase as the offered load increases. Using *1%* outage probability and less than *1%* blocking probability as the system operating point, system capacity is *20* and *10* Erlangs for *V=0.4* and *V=1*, respectively.

Table 4 shows the effect of T_{ADD} on system performance. By decreasing T_{ADD}, the soft handoff region increases which can be seen from the handoff probabilities. Due to this additional diversity benefit, a reduction is observed in the average power allocation as seen in Table 4. This reduction in the power allocation results in an increase in system capacity. For *V=1* a system using a T_{ADD} of *–16* dB can support *10* Erlangs, however a system using a T_{ADD} of *–18* dB can support *12* Erlangs. On the other hand, the average number of traffic channels normalized to the mean offered load increases as T_{ADD} decreases. Thus, putting a burden on the hardware requirement at the cell site. This effect can be seen from Fig. 4 which shows the CDF of mobiles that demand service, i.e. N_h before limiting it to 60, as a function of N_s. Table 4 shows that coverage is almost the same for both cases of T_{ADD}.

5. CONCLUDING REMARKS

A model for analyzing the CDMA forward link performance in a flat fading channel was developed. The effect of power allocation on the system capacity was evaluated. It was shown that the CDMA forward link performance is robust to slow flat fading conditions due to the use of soft handoff. However, soft handoff increases the hardware requirement at the cell site. System capacity is sensitive to combinations of several design parameters, e.g. T_{ADD}, outage, coverage, blocking and the target *FER*. Simulations indicate that a capacity of *10–20* Erlangs can be obtained in a system with omnidirectional antennas.

REFERENCES

[1] K. Gilhousen, et al, "On the Capacity of a Cellular CDMA System ," *IEEE Trans. on Veh. Tech.*, vol. 40, no. 2, pp. 303—331, May 1991.

[2] R. Padovani, B. Butler and R. Bosel, "CDMA Digital Cellular: Field Test Results," *Proc. IEEE Veh. Tech. Conf.*, pp. 11—15, (Stockholm, Sweden), May 1994.

[3] TIA/EIA/IS–95 Interim Standard, Telecommun. Industry Association, July 1993.

[4] R. Bultitude, et al, "Propagation Characteristics on Microcellular Urban Mobile Radio Channels at 910 MHz," *IEEE J. on Sel. Areas in Commun.*, vol. 7, no. 1, pp. 31—39, Jan. 1989.

[5] A. M. Viterbi and A. J. Viterbi, "Erlang Capacity of a Power Controlled CDMA System," *IEEE J. on Sel. Areas in Commun.*, vol. 11, no. 6, Aug. 1993.

[6] R. Padovani, "Reverse Link Performance of IS–95 Based Cellular Systems," *IEEE Personal Commun. Mag.*, pp. 28—34, Third Quarter 1994.

[7] L. Jalloul and J. Holtzman, "Performance Analysis of DS/CDMA with Noncoherent M–ary Orthogonal Modulation in Multipath fading Channels," *IEEE J. on Sel. Areas in Commun.*, vol. 12, no. 5, pp. 862–870, June 1994.

[8] R. Pickholtz, et al, "Spread Spectrum for Mobile Communications," *IEEE Trans. on Veh. Tech.*, vol. 40, no. 2, pp. 323—332, May 1991.

[9] S. Wang and I. Wang, "Effects of Soft Handoff, Frequency Reuse and Non–Ideal Antenna Sectorization on CDMA System Capacity," *Proc. IEEE Veh. Tech. Conf.*, pp. 850–854, 1993.

[10] S. Wang and I. Wang, "Simulation Results on CDMA Forward Link System Capacity," *Proc. WINLAB Workshop*, pp. 227–235, (Rutgers University, New Jersey), October 1993.

[11] A. Jalali and P. Mermerlstein, "Effects of Diversity, Power Control, and Bandwidth on the Capacity of Microcellular CDMA Systems," *IEEE J. on Sel. Areas in Commun.*, vol. 12, no. 5, pp. 952—961,June 1994.

[12] M. Wallace and R. Walton, "CDMA Radio Network Planning," *Proc. Int. Conf. on Universal Personal Commun.*, pp. 62–67, (San Diego, CA), Sept. 1994.

[13] A. J. Viterbi, et al., "Other–Cell Interference in Cellular Power–Controlled CDMA," *IEEE Trans. on Commun.*, vol. 42, No.2/3/4, pp. 1501—1504, Feb./Mar./April 1994.

[14] J. Proakis, *Digital Communication,* Second Edition, McGraw—Hill, 1989.

[15] M. Chopra, K. Rohani and J. D. Reed, "Analysis of CDMA Range Extension due to Soft Handoff," to appear in *IEEE Veh. Tech. Conf.*, July 1995.

Table 4. System performance as a function of T_{ADD}.

Other cell interference	Mean Power (V=0.4)		Full power (V=1)	
T_{ADD} (dB)	–18	–16	–18	–16
Average power allocated (% of total power)	2.1	2.8	4.1	5
No handoff (%)	36	44	46	57
2–way soft handoff (%)	25	30	30	31
3–way soft handoff (%)	39	26	24	12
Mean offered load (Erlangs)	20	20	12	10
Avg. TCH/Erlang	2	1.8	1.8	1.54
Outage Prob. (%)	0.1	0.9	1	0.7
Blocking Prob. (%)	2.3	0.7	0	0
Coverage Prob. (%)	99	99	90	90
Average FER (%)	0.92	0.95	1	1.1

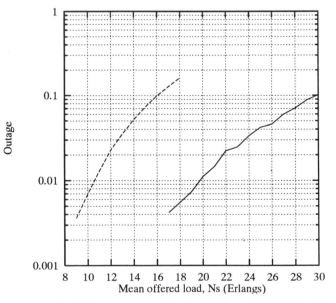

Fig. 2. Outage versus mean offered load for T_{ADD}=–16dB. Solid line for other cells at mean power and dashed line for other cells at full power.

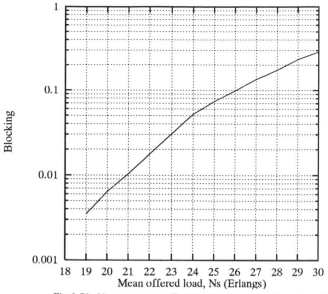

Fig. 3. Blocking versus mean offered load for T_{ADD}=−16dB and other cells at mean power.

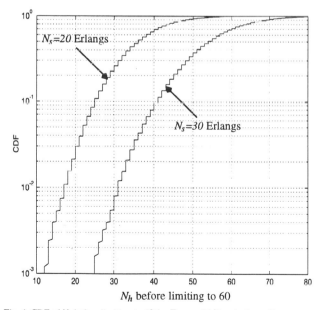

Fig. 4. CDF of N_h before limiting to 60 for T_{ADD}=−16dB and other cells at mean power.

THE EFFECT OF DIRECTIONAL ANTENNAS IN CDMA WIRELESS LOCAL LOOP SYSTEMS

Alper T. Erdogan , Ayman F. Naguib , Arogyaswami Paulraj

Information Systems Laboratory
Stanford University
Stanford, CA 94305

Abstract

The use of directional antennas at the subscriber unit in CDMA wireless local loop systems is studied. Directional antennas can reduce multiuser interference at the CDMA base station and lower subscriber transmit power. The effect of these gains on performance gains in cell capacity and cell coverage are presented. Our analysis focuses on the reverse mobile to base link only.

1. Introduction

Wireless local loop (WLL) systems provide voice and data communication connectivity to Public Switch Telephone Networks (PSTN) via radio links for fixed subscribers. There are many features that make WLL systems attractive over conventional wired local loops. For example, WLL systems are much faster to deploy than wired loops and often offer lower capital costs. One important factor to reduce capital costs is high spectrum efficiency. Larger the number of subscribers supported by the base station, smaller is the share per subscriber of the base station and other infrastructure costs. Since WLL serves a cost sensitive market, the need to improve spectrum efficiency is particularly strong.

In CDMA WLL systems, all cells share the same frequency band and users are separated by different quasi orthogonal spreading codes. Therefore, every user signal appears as interference to every other user. The signals received at the base are correlated by their own spreading codes which results in a spreading gain that boosts E_b/N_o. The system capacity is set by the number of users that can be permitted per cell before the E_b/N_o drops below a threshold set to provide a target BER. It is simple to show [1], that the E_b/N_o with N users per cell, is given by

$$\frac{E_b}{N_o} = \frac{LP_s}{P_n + \alpha.((N-1).P_s + f.N.P_s)} \tag{1}$$

Here P_s is the received user power level and is assumed to be same for all users (perfect power control), P_n is the receiver thermal noise power, L is the processing gain, α is the voice activity factor and f is the ratio of outer cell interference to own cell interference. α is typically around 0.4 and f is around 0.55. The use of antenna arrays at base station in cellular CDMA can reduce co-channel interference and has been studied [2], [3], [4]. Base station arrays form directional beams toward each subscriber therefore reduce interference

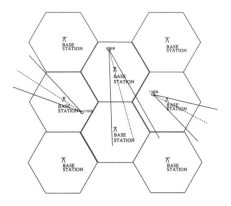

Fig. 1. Subscriber geometry for directional antenna usage

and increase E_b/N_o and can therefore be used to increase capacity.

In this paper, we study the use of directional antennas at the subscriber unit. This is feasible in WLLs since the subscribers are stationary and also the antennas are likely to be on elevated roof tops and can therefore have a path relatively obstructed by local structures. In contrast, this is not easily possible in mobile applications, where the mobile is constantly maneuvering and the path to the base is usually obstructed by buildings around the mobile. We assume that each subscriber antenna in the WLL is pointed towards the parent base station as shown in Figure 1. Such subscriber antennas can improve both reverse and forward link performance. In this paper, we will investigate only the reverse link benefits. Similar advantages should be possible on the forward link also. Directional subscriber antennas reduce the effect of outer cell interference and therefore improve capacity. Besides, there are advantages of reduced transmit power from the subscriber which may not be an important motivation except in rural applications where solar or battery power is used.

In Section   we study the effects of using subscriber unit directional antennas, Section discusses capacity and transmit power improvement as a result of directional antenna usage, in Section   the range increase is analyzed for the cases of constant number and constant density of users, and finally in conclusion, Section 0-B, we present a summary of the paper.

2. Directional Subscriber Antennas

Directional subscriber antennas reduce effect of outer cell interference since each subscriber directs his energy mainly to the (desired) base station and sends only low sidelobe radiation to the other base station. If we assume that each subscriber is linked to the closest base station, then only subscribers in certain areas of the outer cells will generate interference at a reference base. These areas are shown in Figure 2, for a beamwidth of 45°. However, if the subscribers are assigned to the base with the least path loss, then due to variable shadowing effects, subscribers may some time be assigned to a base station other than the closest base.

Fig. 2. Regions of the outer cells that can create interference for subscriber antenna beamwidth of 45° with nearest cell assignment method.

In our model each subscriber is assigned to the base station with the least path loss among the four closest base stations where, as suggested in [5], the path loss from a user to a base station combines spreading loss and lognormal shadowing and is given by

$$\alpha_i(r, \zeta) = r_i^4 10^{\zeta_i/10} \qquad i = 1..4 \qquad (2)$$

where r_i is the distance to base station, ζ_i is a normal variate with zero mean and standard deviation σ. In our simulations, σ is chosen to be equal to 8 dB. Furthermore, ζ_i is defined as weighted sum of two normal random variables ξ and ξ_i, both with zero means and standard

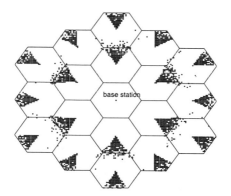

Fig. 3. Regions of the outer cells that can create interference for subscriber antenna beamwidth of 45° with the least path loss assignment method.

74

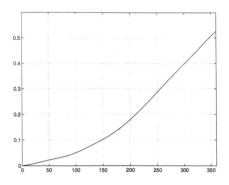

Fig. 4. Ratio of outer to inner cell interference, f, as a function of beamwidth.

deviations $\sigma = 8$. *i.e.* :

$$\zeta_i = a\xi + b\xi_i \tag{3}$$

where $a^2 + b^2 = 1$ and we have chosen $a^2 = b^2 = 1/2$. Here ξ represents near field component of the log normal fading and is same for all paths from a subscriber to base stations and ξ_i is the log normal component specific to i^{th} path. With the above assignment model, the out of cell subscribers that interfere with the reference base station are shown in Figure 3. The interference generating locations are similar to the that in Figure 2, but with some randomization due to effect of shadowing.

As pointed earlier the main effect of subscriber antennas is to reduce out of cell interference which is given by the f factor. In order to determine how f depends on the antenna beamwidth, simulations were carried out using path loss model in Eq 2 and assuming perfect power control and approximating of the beam response by a brick wall pattern. The results are given in Figure 4. We show that f increases with antenna beamwidth starting with $f = 0$ at zero beamwidth and reaching a value of $f = 0.53$ for omni directional antenna.

3. Improvements in Network Capacity

The capacity improvement of a cell is an important benefit obtained from the reduction in the outer cell interference. An expression for capacity in terms of the number of users in a cell, N, can be obtained by rewriting Eq 1 as

$$N = \frac{P_s/P_n.(L + \alpha.E_b/N_o) - E_b/N_o}{\alpha.E_b/N_0.(1 + f).P_s/P_n} \tag{4}$$

This capacity is maximized by making the received user power infinite and therefore thermal noise power is negligible. This yields the pole capacity and can be found by letting P_s go to ∞

$$N_\infty = \frac{L + \alpha.E_b/N_o}{\alpha.E_b/N_o.(1 + f)} \tag{5}$$

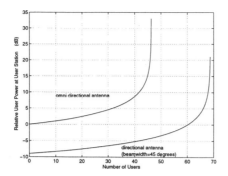

Fig. 5. Rise Curve: Transmit power at mobile vs number of users (relative to power of single user with omni directional antenna

From Eq 5, we can see that the pole capacity will increase as the f reduces. As f approaches zero, the increase in pole capacity will be approximately 55 percent.

It is also interesting to see if a reduction in f has any influence on required mobile power received at the base. Again by rewriting Equation 1, an expression for the received power at the base station is obtained as

$$P_s = \frac{E_b/N_o.P_n}{L + \alpha.E_b/N_o.(1 - (1 + f).N)} \tag{6}$$

It is apparent from the above that the required received power at the base with directional antennas is less than that needed with omni directional antennas for the same cell loading. This implies that directional antennas reduce demand for subscriber power, and this reduction arises both from reduced interference and antenna gain.

These improvements in cell capacity and subscriber power are shown in Figure 5 where two "rise curves" one for omni directional antennas and the other for directional (beamwidth 45°) are shown. The rise curve shows the relationship between the transmitted power by the subscriber (sometimes also plotted for the received power at the base) as a function of cell loading. The rise curve starts at its lowest point for a single user and increases gradually till it reaches a vertical asymptote (pole) at maximum loading. The rise curve for the directional antennas starts at a lower power by an amount corresponding to the array gain and reaches the pole at a greater cell loading due to reduced outer cell interference.

Since the path loss is strongly dependent on the range, it is clear that the subscribers close to the base station have lower transmitted power as compared to those at the cell edge. Therefore, the outer cell interference potential of subscribers near their own cell edge will be higher. Also, since higher the gain of an antenna, more its cost, a cost effective approach to using subscriber antennas is to use an antenna beamwidth (or array gain) proportional to the needed EIRP or equivalently inversely proportional to the average path loss. The path loss is a sum of range dependent loss and shadowing loss and therefore has a random component. To use a simple and a more feasible strategy for subscriber antenna gain, we

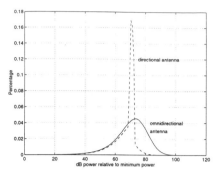

Fig. 6. Power histograms for omnidirectional and variable gain antenna assignment cases. Absicca is scaled relative to minimum power.

assigned antennas beamwidths from a selection of 5 gains : 12, 9, 6, 3, 0 dB. The assignment was based on the EIRP required to meet received power at the base stations. The highest power subscribers were assigned the 12 dB antennas and the lower power subscribers are assigned lower gain antennas based on equal interference principle. To see the effect of such variable gain antennas we studied the histogram of subscriber power with uniform omni directional antennas and variable gain antennas. The results are shown in Figure 6 where the use of variable gain antennas reduce the variability of required user power. The use of variable gain antennas as described above reduced the f factor from 0.53 to 0.16.

4. Improvements in Cell Coverage

Another benefit that can be obtained from using subscriber directional antennas is the increase in range for constant subscriber transmit power. While the increased array gain

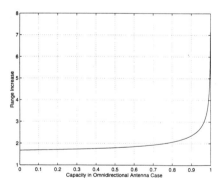

Fig. 7. Range improvement vs loading for 45° subscriber antenna beamwidth (constant number of users)

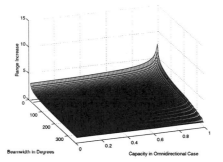

Fig. 8. Effect of beamwidth on the increase in range for constant number of users.

boosts EIRP and therefore increases array gain, the reduced outer cell interference also plays an additional role in increasing range performance. This effect is discussed for two different cases:

A. Increase In Range For Constant Number Of Users

The number of users within the cell is constant and we can show that the effect of directional antennas on range improvement is given by

$$m = (G_a(1 + \alpha(\frac{P_s}{P_n})(f_1 - f_2)N))^{1/4} \tag{7}$$

where G_a is the directional antenna gain, f_1 and f_2 represent outer cell interference factors for omni directional and directional antennas respectively.

The first issue to examine is the effect of base cell loading on range improvement. Since the range depends on transmit (and therefore received) power. And further from the rise curve the received power is function of cell loading, the range increase with directional antennas

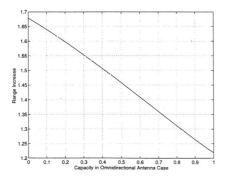

Fig. 9. Range increase vs loading for 45° subscriber antenna beamwidth (constant density of users)

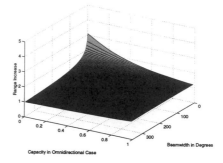

Fig. 10. Effect of beamwidth on the increase in range for constant density of users

will depend on the cell loading. This is illustrated in Figure 7 for a 45° antenna and shows that as the percentage cell loading (referenced to omnidirectional antenna) increases, the *relative* range improvement increases. As we approach the pole capacity, the powers go to infinity and the corresponding range increase factor also becomes unbounded.

In Figure 8 we plot the effect of different beamwidths as a function of cell loading. We can see that smaller beamwidths provide higher range increase due to reduction in f_2 factor. Also, this effect becomes more pronounced at higher cell loadings.

B. Increase In Range For Constant Density of Users

We now re-examine range improvement with constant user areal density model. In this model increasing cell size also means increased number of users in the cell. Adding new users will increase interference and therefore tends to reduce range increases.

The increase in range as a function of cell loading with beamwidths of 45° shows very different behavior than the constant number of users case. The results are given in Figure 9. For this case, range increase is inversely proportional to initial omnidirectional capacity, because the system is more sensitive to addition of new users and therefore also their interference at high capacity loadings.

The effect of different beamwidths on the range increases is shown in in Figure 10. Just as in the case of constant number of users, range improves with smaller beamwidths.

5. Conclusion

The use of directional antennas at the subscribers in wireless local loop systems increase cell capacity and range coverage. Cell capacity improvements can be largely preserved by using variable gain antennas where the higher power cell edge subscribers use higher gain antennas than closer in subscribers. Range improvements in general result from reduced antenna beamwidths. However, the percentage improvements depend on the subscriber density models and the cell loading levels.

We believe these results show that use of variable gain antennas can be very useful in improving network economics in wireless local loop systems.

REFERENCES

[1] A. M. Viterbi and A. J. Viterbi, "Erlang Capacity of a Power Controlled CDMA System," *IEEE J. Select. Areas Commun.*, vol. 11-(6), pp. 892–900, August 1993.

[2] S. Simanapalli, "Adaptive Array Methods for Mobile Communications," in *Proc. VTC'94*, vol. II, (Stockholm, Sweden), June 1994.

[3] S. C. Swales, M. A. Beach, D. J. Edwards, and J. P. McGeehn, "The Performance Enhancement of Multibeam Adaptive Base Station Antennas for Cellular Land Mobile Radio Systems," *IEEE Trans. Veh. Tech.*, vol. VT-39(1), pp. 56–67, February 1990.

[4] A. F. Naguib, A. Paulraj, and T. Kailath, "Capacity Improvement with Base-Station Antenna Arrays in Cellular CDMA," *IEEE Trans. Veh. Tech.*, vol. VT-43(3), pp. 691–698, August 1994.

[5] A. M. Viterbi, A. J. Viterbi, and E. Zehavi, "Other Cell Interference in Cellular Power-Controlled CDMA," *IEEE Trans. Commun.*, vol. 42-(4), pp. 1501–1504, April 1994.

Decision-Directed Coherent Delay-Locked PN Tracking Loop

for DS-CDMA

M. SAWAHASHI and F. ADACHI

R&D Department, NTT Mobile Communications Network Inc.

1-2356 Take, Yokosuka-shi, Kanagawa-ken, 238-03 Japan

Phone : +81-468-59-3487, Fax : +81-468-57-7907

Abstract

In this paper, a new implementation of the decision-directed coherent PN delay locked loop (DLL) is proposed for DS-CDMA that uses pilot symbol-aided coherent data detection. The performance of the proposed coherent DLL is evaluated by computer simulation. Computer simulation results show that it can significantly reduce the rms tracking jitter of the regenerated spreading code replica, thereby improving bit error rate (BER) performance in fading environments.

1.Introduction

Recently, there has been increasing interest in applying DS-CDMA applications to cellular mobile radio systems[1],[2]. Fast and accurate spreading code acquisition and high precision tracking are essential. The noncoherent delay-locked tracking loop (DLL) is the most popular and well developed technique [3]. Although it provides a 3dB increase in the signal-to-noise ratio of the code timing error signal which is obtained by taking the cross correlation between received signal and the regenerated spreading code replica, compared to an alternate time-shared method such as the tau dither loop (TDL) [4], it suffers from increased tracking jitter due to the noise enhancement (squaring loss) arising from the square-law detector. The decision-directed coherent DLL [5],[6] overcomes this problem. The performance of the decision-directed coherent DLL is analyzed theoretically in [6] assuming additive white Gaussian (AWGN) channels. It requires accurate estimation of the received carrier phase, which is difficult in

fading environments. Furthermore, no practical implementation was suggested in [6]. To increase the capacity of DS-CDMA, coherent data detection is most attractive. Since in fading environments the received carrier phase rapidly changes in random manner, the pilot symbol-aided interpolation technique [7] can be applied. It estimates the fading-induced carrier phase from the periodically inserted known pilot symbols for coherent data detection. The estimated reference carrier can be also used to realize a coherent DLL.

This paper proposes the practical implementation and linear analysis of the decision-directed coherent DLL. Theoretical analyses of rms tracking jitter of the decision-directed coherent DLL and of the conventional noncoherent DLL are then presented. Next, the computer simulation results for the root mean square of the timing error (rms tracking jitter) of the regenerated spreading code replica are presented. Finally, the BER performance achievable with coherent DLL and pilot-symbol coherent detection is evaluated.

2. Decision-directed coherent DLL

Fig.1 shows the block diagram of the proposed decision-directed coherent DLL. First, we describe the data detection process. The received spread signal is transformed into the complex spread signal, composed of in-phase(I) and quadrature-phase(Q) components, by the quasi-coherent quadrature detector and is oversampled at a rate more than double the chip rate. This complex spread signal is then multiplied with the locally regenerated spreading code replica $c(t-\hat{\tau})$, where $\hat{\tau}$ is the estimate of the channel delay, and integrated during one data symbol duration T to obtain a narrowband modulated signal sample. Coherent data detection using pilot symbol interpolation is then applied; the reference phase for coherent detection is estimated by interpolating the carrier phases associated with the known pilot symbols and is also used for coherent DLL.

The proposed coherent DLL operates as follows. The complex spread signal is cross-correlated with advanced and retarded code replicas, $c(t-\hat{\tau}+\Delta)$ and $c(t-\hat{\tau}-\Delta)$, in the DLL branches, where Δ is the chip time offset. The integrate-and-dump (I & D) filter outputs at $t=iT$ of two DLL branches are combined to obtain the chip timing error signal ε_i which contains the data modulation and the carrier phase and amplitude variations due to fading. The chip timing error signal ε_i is reverse-modulated by feeding back the detected data $\hat{d}_i$ and the estimated fading envelope $\hat{\xi}_i$ to obtain the chip timing error $\hat{\varepsilon}_i$. The resulting chip timing error is smoothed by the loop filter to control the numerical controlled code generator (NCCG).

3. Linear analysis of decision-directed coherent DLL

The complex spread signal is represented as

$$r(t) = \sqrt{2P}\,d(t-\tau)c(t-\tau)\,\xi(t) + w(t) \quad (1)$$

where $c(t)$ is the spreading waveform, $d(t)$ is the data-modulated signal waveform, $\xi(t)$ is the time varying complex fading envelope, τ is the time delay incurred by the channel, and $w(t)$ is the receiver composite noise with power spectral density N_0 which is the sum of the additive white Gaussian noise (AWGN) and other-user interference. We approximate $w(t)$ as the complex Gaussian noise. In Eq. (1), $c(t)$ and $d(t)$ are expressed as

$$c(t)=\sum_{i=-\infty}^{\infty}\sum_{k=0}^{pg-1} c_k\, p\!\left(\frac{t}{T_c} - ipg-k\right), \quad d(t)=\sum_{i=-\infty}^{+\infty} d_i\, p\!\left(\frac{t}{T} -i\right), (2)$$

where $\{c_k\}$ and $\{d_i\}$ are the binary chip sequence with repetition period of pg chips and the transmitted data sequence, respectively, T_c and T are the chip duration and modulation symbol duration, respectively, $pg=T/T_c$, and $p(t)=1$ (0) if $0\leq t\leq 1$ (otherwise). $r(t)$ is cross-correlated with the advanced and retarded spreading code replicas, $c(t-\hat{\tau}+\Delta)$ and $c(t-\hat{\tau}-\Delta)$, to obtain the chip timing error signal. The chip timing error signal sampled at $t=iT$ is given by

$$\varepsilon_i=\frac{1}{T}\int_{iT}^{(i+1)T} r(t)\big[c(t-\hat{\tau}-\Delta) - c(t-\hat{\tau}+\Delta)\big]dt \quad . \quad (3)$$

For large pg, the effects of the time delay τ at the beginning and ending of the integration of Eq.(3) can be ignored, so ε_i can be approximated as

$$\varepsilon_i=\sqrt{2P}\,d_i\,\xi_i\left[R_c(-\Delta+\delta T_c)-R_c(\Delta+\delta T_c)\right] + w_i \quad , \quad (4)$$

where $\xi_i = \xi(iT)$, $\delta = (\tau-\hat{\tau})/T_c$ is the normalized

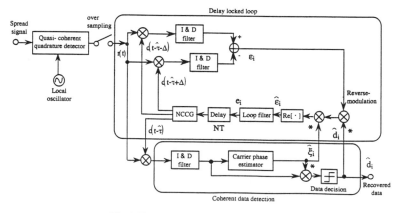

Fig.1 Decision-directed coherent DLL.

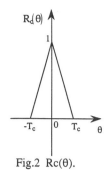

Fig.2 Rc(θ).

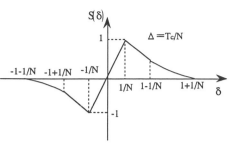

Fig.3 Discriminator characteristics.

chip timing error, and $R_c(\theta)$ is the auto-correlation function of the spreading code defined as

$$R_c(\theta) = \frac{1}{pgT_c} \int_0^{pgT_c} d(t)d(t+\theta)dt \quad . \tag{5}$$

The $R_c(\theta)$ assumed for our analysis is shown in Fig.2. In Eq.(4), w_i is the complex noise given by

$$w_i = \frac{1}{T} \int_{iT}^{(i+1)T} w(t)\left[d(t-\hat{\tau}-\Delta) - d(t-\hat{\tau}+\Delta)\right] dt \tag{6}$$

and its variance is given by

$$E\left[|w_i|^2\right] = \frac{4N_0}{T}\left[1 - R_d(2\Delta)\right] , \tag{7}$$

where $E[\cdot]$ denotes ensemble average. The chip timing error signal ε_i contains the complex fading envelope ξ_i and the data modulation d_i, both of which should be removed by reverse-modulation. ξ_i is estimated by pilot symbol interpolation (the estimate of ξ_i is denoted as $\hat{\xi}_i$). The sequence of ε_i is reverse-modulated to obtain the sequence of the corrected chip timing error signal $\hat{\varepsilon}_i$, which is expressed

$$\hat{\varepsilon}_i = \mathrm{Re}\left[\varepsilon_i\left(\hat{d}_i\frac{\hat{\xi}_i}{|\hat{\xi}_i|}\right)^*\right] , \tag{8}$$

where $\mathrm{Re}[z]$ denotes the real part of z and $*$ represents the complex conjugate. Substituting Eq.(4) into (8), we obtain

$$\hat{\varepsilon}_i = \mathrm{Re}\left[\sqrt{2P}d_i\left(\hat{d}_i\frac{\hat{\xi}_i}{|\hat{\xi}_i|}\right)^*\left[R_c(-\Delta+\delta T_c)-R_c(\Delta+\delta T_c)\right]\right]$$
$$+ \mathrm{Re}\left[w_i\left(\hat{d}_i\frac{\hat{\xi}_i}{|\hat{\xi}_i|}\right)^*\right] . \tag{9}$$

The sequence of $\hat{\varepsilon}_i$ is fed to the digital loop filter to obtain the NCCG control signal. The first

term of Eq.(9) contains the desired chip timing error and the second is the noise due to AWGN plus other-user interference. DS-CDMA is operated at low signal-to-AWGN plus interference ratios, so we approximate the first term using the average $M(\delta) = E\left[\hat{d}_id_i\right]$ and ignore the modulation self-noise term $\hat{d}_id_i - E\left[\hat{d}_id_i\right]$ with negligible effect. We also ignore the effect of estimation error in $\hat{\xi}_i$ and assume $\hat{\xi}_i = \xi_i$ (instead, the estimation error effect is included in the probability of decision error). Denoting the probability of decision error at the normalized chip timing error δ by $Pe(\delta)$, $M(\delta)$ is given by

$$M(\delta) = \left[1-2P_e(\delta)\right] . \tag{10}$$

Denoting the impulse response of the digital loop filter by g_l , $l = 0,1,2, \cdots$, the filter output c_i which controls the NCCG can be expressed as

$$c_i = \mathrm{Re}\left[\hat{\varepsilon}_i\otimes g_l\right] = \sqrt{2P}|\xi_i|\eta M(0)\left[S(\delta) + \frac{N_i}{\sqrt{2P}|\xi_i|\eta M(0)}\right]\otimes g_l \tag{11}$$

where $\otimes$ denotes convolution and $S(\delta)$ is the discriminator characteristics (S-curve). $S(\delta)$ and η are given by

$$S(\delta) = \frac{M(\delta)}{\eta M(0)}\left[R_c(-\Delta+\delta T_c)-R_c(\Delta+\delta T_c)\right]$$
$$\eta = \frac{\partial}{\partial\delta}\left[R_c(-\Delta+\delta T_c)-R_c(\Delta+\delta T_c)\right]|_{\delta=0} . \tag{12}$$

$$N_i = \mathrm{Re}\left[w_i\left(\hat{d}_i\frac{\hat{\xi}_i}{|\hat{\xi}_i|}\right)^*\right]$$

An example of $S(\delta)$ when $\Delta = T_c/N$ is shown in Fig.3 (we ignored the effect of probability of decision error, i.e., $M(\delta) = M(0)$). The linear S-curve is obtained for $|\delta| \leq \Delta/T_c$. For our time discrete representation, the stochastic differential

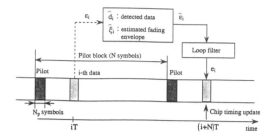

Fig.4 Operational timing of decision-directed coherent DLL.

equation of DLL operation [6] is given by $Ke_i = -(\Delta\delta_i/T)$, where K denotes the loop gain, δ_i is the value of δ at $t=iT$, and $\Delta\delta_i = \delta_{i+1} - \delta_i$. Since $\eta=2$ and $S(\delta_i)=\delta_i$ in the linear region ($|\delta_i| < \Delta/T_c$), we obtain using Eq.(11)

$$\frac{\Delta\delta_i}{T} = -2\sqrt{2P}|\xi_i|KM(0)\left[\delta_i + \frac{N_i}{2\sqrt{2P}|\xi_i|M(0)}\right]\otimes g_l. \quad (13)$$

Equation (13) represents the operation of coherent DLL in Rayleigh fading environments. Since the pilot symbol-aided coherent detection is block coherent detection, the sequences of $\hat{\xi}_i$ and $\hat{d}_i$ are obtained at the end of the block. Hence, the loop filter output ε_i associated with time position iT is used for updating the timing of the regenerated spreading code replica at time $t=(i+N)T$, where N is the block length (see Fig.4).

4.Analysis of tracking jitter

Since it is not easy to analyze the tracking jitter in fading environments because of the varying noise power of the second term inside the square blackets in Eq.(13), we assume non fading environments ($\xi_i=1$). Assuming a large loop gain K, the transfer function of DLL can be approximated by that of the loop filter. In our analysis, we assume that the loop filter is a moving average filter that averages the past m samples. The variance σ_δ^2 of the normalized chip timing error δ_i is that of the second term inside the square bracket of Eq.(13) [8]. σ_δ^2 is given by

$$\sigma_\delta^2 \approx \frac{1}{8PM^2(0)}\frac{E[N_i^2]}{m} \quad (14)$$

since N_i's are independent, zero-mean Gaussian noise. From Eq.(7),

$$E[N_i^2] = \frac{E[|w_i|^2]}{2} = \left(\frac{2N_0}{T}\right)[1-R_c(2\Delta)] \quad (15)$$

and we obtain

$$\sigma_\delta^2 \approx \frac{1-R_c(2\Delta)}{4\frac{E_s}{N_0}M^2(0)m^2}, \quad (16)$$

where $E_s=PT$ is the signal energy per modulation symbol. For BPSK data modulation ($E_s/N_0=E_b/N_0$), the probability of decision error is given by

$$P_e(0)=Q\left(\sqrt{2\frac{E_b}{N_0}D^{-1}}\right), \quad (17)$$

where D denotes the noise enhancement due to pilot symbol interpolation and

$$Q(x)=\frac{1}{\sqrt{2\pi}}\int_x^\infty \exp(-y^2/2)dy. \quad (18)$$

D is, from [7], obtained as

$$D=1+\frac{\sigma^2}{\left(\frac{2N_0}{T}\right)}, \quad (19)$$

where σ^2 is the variance, averaged over N-Np data positions, of the noise component of the estimated reference signal $\hat{\xi}_i$. For the linear interpolation using N_p-symbol pilots and N-symbol pilot blocks, σ^2 can be given by

$$\sigma^2=\frac{(2N_0/T)}{N_p(N-N_p)}\sum_{k=1}^{N-N_p}\left[1-2\left(\frac{(N_p-1)/2+k}{N}\right)+2\left(\frac{(N_p-1)/2+k}{N}\right)^2\right] \quad (20)$$

for Np=even. Consequently, the rms tracking jitter becomes

$$\sigma_\delta=\frac{1-R_c(2\Delta)}{2m\sqrt{\frac{E_b}{N_0}\left[1-2Q\left(\sqrt{2\frac{E_b}{N_0}D^{-1}}\right)\right]}}$$

for coherent DLL. $\quad (21)$

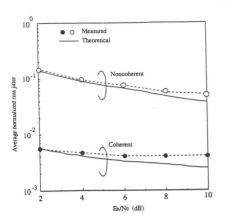

Fig.5 Rms tracking jitter in no fading channel.

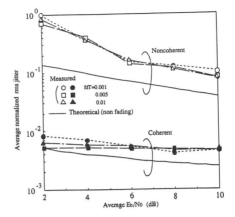

Fig.7 Avarage rms tracking jitter in reverce link.

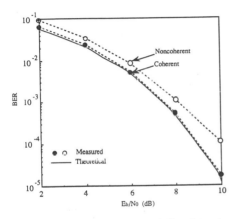

Fig.6 BER performance in no fading channel.

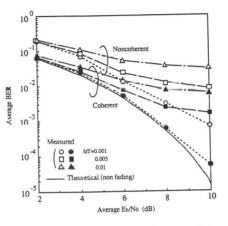

Fig.8 Avarage BER performance in reverce link.

For comparison, we also analyzed the rms tracking jitter of the noncoherent DLL in non fading environments. From the Appendix, we obtain

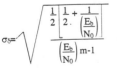

$$\sigma_\delta = \sqrt{\frac{\frac{1}{2}\left[\frac{1}{2} + \frac{1}{\left(\frac{E_b}{N_0}\right)}\right]}{\left(\frac{E_b}{N_0}\right)m\text{-}1}}$$

for noncoherent DLL.   (22)

5.Simulation results

We applyied the proposed decision-directed coherent DLL to the DS-CDMA reverse link and forward link and measured the average rms tracking jitter and the average BER. The absence of a squaring detector in the proposed coherent DLL reduces the steady state rms tracking jitter, thereby improving the BER. Both the data modulation (1st modulation) and spreading (2nd modulation) schemes employ BPSK. The spreading code used here is the Gold sequence with a repetition of 31 chips. For pilot symbol-aided coherent data detection, a known two-symbol pilot is periodically inserted every 14 data symbols ($N=16$, $N_p=2$). In this case, the noise enhancement D becomes D=1.28dB. The chip time offset Δ in the DLL loop was one half the chip duration ($\Delta = T_c/2$). The loop filter considered here computes the sum of the signs of the last m chip timing error signals $\hat{\varepsilon}_i$'s. If the loop filter output is larger than K (less than -K), the NCCG advances (retards) the regenerated spreading code replicas by Δ. In the simulations, $m=20$ and $K=14$ are used.

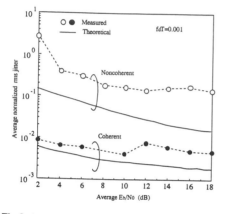

Fig.9 Avarage rms tracking jitter in forward link.

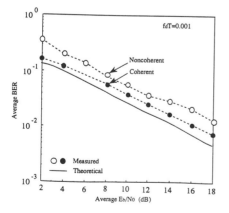

Fig.10 Avarage BER performance in forward link.

5.1 Non fading environment

The normalized rms tracking jitter versus E_b/N_0 is shown in Fig.5. The theoretical values calculated from Eq.(21) and (22) are also shown. The measured results are close to the theoretical curves. The rms tracking jitter is reduced by more than one order of magnitude compared to the conventional, noncoherent DLL. Fig.6 compares the BER performances achievable by coherent and noncoherent DLL. The E_b/N_0 required for BER = 10^{-2} (10^{-4}) is reduced by 0.8dB (1.2dB) compared to the noncoherent DLL and the coherent DLL achieves the BER performance close to the theoretically predicted BER with pilot symbol-aided coherent detection. Since pilot symbol-aided coherent detection is block detection, the chip timing error sequence is delayed by one pilot period. However, the insertion of this delay does not degrade the tracking performance .

5.2 Reverse link channel

For the reverse link, precise power control is necessary so that all user-transmitted signals are received with the same power in order to avoid the near/far problem. Perfect transmission power control was assumed; the amplitude of the received signal is kept constant and only phase variation due to fading remains. The measured average rms tracking jitter is shown in Fig.7 for $f_dT=10^{-2}$, 5×10^{-3}, and 10^{-3}, where f_d is the maximum fading Doppler frequency. Fig.8 shows the average BER performance as a function of average E_b/N_0 (note that N_0 is the spectrum density of AWGN plus aggregate other-user interference). The coherent DLL reduces the required E_b/N_0 at an average BER of 1×10^{-2} by about 1.4dB compared to the conventional DLL. For slow fading ($f_dT=10^{-3}$), almost the same BER performance as the theoretical prediction in non fading channels is obtained, since the fading induced random phase is almost completely suppressed by pilot symbol interpolation.

5.3 Forward link channel

In the forward link, the same-cell other-user interference suffers the same fading as the desired signal. The average signal energy per bit-to-the same-cell other-user interference power spectral density ratio (E_b/I_0) was set to 8dB; this corresponds to 5 users simultaneously communicating in the cell site. In the simulation, we assumed no transmit power control and that the other-cell other-user interference is approximated as an independent, zero-mean Gaussian variable. The measured rms tracking jitter and average BER performance are shown in Figs.9 and 10, respectively, for $f_dT=10^{-3}$ as a function of E_b/N_0 (in this case, note that N_0 is the spectrum density of AWGN plus other-cell other-user interference). Similarly to the reverse link, about 1dB improvement in the required E_b/N_0 can be seen.

6. Conclusion

A practical implementation of the decision-directed coherent DLL scheme with pilot symbol-aided coherent detection was proposed. An approximate analysis of the rms tracking jitter of the proposed decision-directed coherent DLL was presented. The tracking jitter and BER performance achievable by the coherent DLL were evaluated by computer simulations for both reverse and forward links of cellular DS-CDMA. Simulation results showed that coherent DLL reduces the rms tracking jitter by almost one order of magnitude, and improves the BER performance by about 1dB at BER=10^{-2}, compared to noncoherent DLL.

APPENDIX-rms tracking jitter of noncoherent DLL

The noncoherent DLL uses square-law envelope detectors, and the chip timing error signal ε_i is given by

$$\hat{\varepsilon}_i = |\varepsilon_{i-}|^2 - |\varepsilon_{i+}|^2$$
$$= 2P|d_i|^2|\xi_i|^2\left[R_c^2(-\Delta+\delta T_c)-R_c^2(\Delta+\delta T_c)\right]$$
$$+ |w_{-\Delta,i}|^2 - |w_{+\Delta,i}|^2$$
$$+2\sqrt{2P}\text{Re}\{d_i\xi_i\left[R_c(-\Delta+\delta T_c)w_{-\Delta,i}^* - R_c(\Delta+\delta T_c)w_{+\Delta,i}^*\right]\} \quad , \quad (A1)$$

where

$$w_{\pm\Delta,i} = \frac{1}{T}\int_{iT}^{(i+1)T} w(t)c\left(t-\hat{\tau}\pm\Delta\right)dt \quad . \quad (A2)$$

Eq.(A1) can be rewritten as

$$\hat{\varepsilon}_i - 2P|\xi_i|^2\left[R_c^2(-\Delta+\delta T_c)-R_c^2(\Delta+\delta T_c)\right]+N_i \quad (A3)$$

since $|d_i|^2=1$, where

$$N_i=|w_{-\Delta,i}|^2 - |w_{+\Delta,i}|^2$$
$$+2\sqrt{2P}\text{Re}\{d_i\xi_i\left[R_c(-\Delta+\delta T_c)w_{-\Delta,i}^* - R_c(\Delta+\delta T_c)w_{+\Delta,i}^*\right]\} \quad . \quad (A4)$$

The first term of Eq. (A3) represents the desired chip timing error signal and the second is the noise due to AWGN and other-user interference. Similarly to the coherent DLL analysis, we obtain

$$e_i = 2P|\xi_i|^2\eta\left[S(\delta) + \frac{N_i}{2P|\xi_i|^2\eta}\right]\otimes g_l \quad (A5)$$

where

$$S(\delta)=\frac{1}{\eta}\left[R_c^2(-\Delta+\delta T_c)-R_c^2(\Delta+\delta T_c)\right] = \delta \quad \text{for } |\delta|\le\frac{\Delta}{T_c}$$

$$\eta=\frac{\partial}{\partial\delta}\left[R_c^2(-\Delta+\delta T_c)-R_c^2(\Delta+\delta T_c)\right]|_{\delta=0} = 4\left(1-\frac{\Delta}{T_c}\right) \quad (A6)$$

Assuming non fading channel ($\xi_i=1$), we have

$$\sigma_\delta^2 \approx \frac{1}{(2P)^2\eta^2}\frac{E[N_i^2]}{m} \quad . \quad (A7)$$

We can show that when $\Delta=T_c/2$, $w_{\pm\Delta,i}$ are independent Gaussian variables with variance of $2N_0/T$ and we obtain, after some calculation,

$$E[N_i^2]=8\left(\frac{N_0}{T}\right)^2 + 8P\left(\frac{N_0}{T}\right)E[R_c^2(-\Delta+\delta T_c)+R_c^2(\Delta+\delta T_c)] \quad (A8)$$

When $\Delta=T_c/2$,

$$E[R_c^2(-\Delta+\delta T_c)+R_c^2(\Delta+\delta T_c)]=2\sigma_\delta^2+\eta^2/8 \quad (A9)$$

Substituting Eqs.(A8) and (A9) into (A7) yields

$$\sigma_\delta=\sqrt{\frac{\frac{1}{2}\left[\frac{1}{2}+\frac{1}{\left(\frac{E_h}{N_0}\right)}\right]}{\left(\frac{E_b}{N_0}\right)m-1}} \quad (A10)$$

for BPSK.

References
[1] A. Salmasi and K. S. Gilhousen, "On the system design aspects of code division multiple access(CDMA) applied to digital cellular and personal communications networks," Proc.VTC'91, pp.57-62.
[2] K. S. Gilhousen, I. M. Jacobs, R. Padovani, A. J. Viterbi, L. A. Weaver, and C. E. Wheatley III, "On the capacity of a cellular CDMA system," IEEE Trans. Veh. Technol., vol. VT-40, No.5, pp.303-312, May 1991.

[3] J. J. Spiker. Jr., "Delay-Lock Tracking of Binary Signals," IEEE Trans. Space Electr. and Telem., vol. SET-9, No.1, pp.1-8, March 1963.

[4] H. P. Hartmann, "Analysis of a Dithering Loop for PN Code Tracking," IEEE Trans. Aerospace Electron. Syst., vol. AES-10, pp.2-9, Jan. 1974.

[5] J. K. Holmes, "Coherent Spread Spectrum Systems" John Wiley and Sons, 1982.

[6] R. D. Gaudenzi and M. Luise, "Decision-Directed Coherent Delay-Lock Loop for DS-Spread-Spectrum Signals," IEEE Trans. Commun., vol. COM-39, No.5, pp.758-765, May 1991.

[7] S. Sampei and T. Sunaga , " Rayleigh Fading Compensation for QAM in Land Mobile Radio Communications, " IEEE Trans. Veh. Technol., vol. VT-42, No.2, pp.137-146, May 1993.

[8] W.C.Lindsey, "Synchronization systems in Communication and Control" Prentice-Hall, Inc., 1972.

Adaptive Beamforming for Wireless Communications

J. Litva, A. Sandhu, K. Cho and T. Lo

Communications Research Laboratory, McMaster University

1280 Main Street West, Hamilton, Ontario, Canada L8S 4K1

1 Introduction

The increasing demand for mobile and cellular services has resulted in the near-depletion of allocated frequency resources, especially in urban areas. In order to absorb more and more customers in a wireless communication system, we face the challenge of improving the system capacity without degradation in the quality of services. Moreover, the information also being transmitted in the form of data, besides voice, calls for a higher system reliability.

The frequency reuse techniques utilize the available frequency spectrum in an efficient manner. In order to increase this frequency reuse factor, the cell size must be reduced. Therefore, mechanisms must be provided to combat co-channel interference and multipath fading effects. One of such methods employs adaptive antennas that perform spatial filtering in order to distinguish between the desired signals and the interfering signals. Alternately, we can use diversity techniques which also make use of antenna arrays. Both of these methods are expected to improve the system capacity within the allocated frequency spectrum.

In this paper, we present two techniques, namely, the use of beamforming and diversity techniques, for practical wireless communication systems. One of the diversity techniques, called the selection diversity technique, has been widely adopted for actual mobile radio systems because of the performance that it delivers with relatively simple hardware requirements. We present the BER performance comparison between the diversity and the beamforming receivers in the presence of interfering signals, and show the advantage of using a beamforming receiver. One disadvantage of the commonly used beamforming algorithm, the Least Mean Square (LMS) algorithm, is its slow convergence rate. In order to overcome this problem, we present a faster neural beamforming technique that can be used in a spread spectrum communication system.

The next section discusses the diversity techniques and compares their BER performance to that of a Least Mean Square (LMS) beamformer. A reference generating technique based on decision feedback is also discussed. In section 3, we present a faster algorithm for adaptive beamforming using neural networks. Also, the application of chaos to spread spectrum communications is discussed. Finally, we present some concluding remarks on the use of diversity and beamforming techniques in wireless communications in section 4.

2 BER Comparison Between Diversity and Beamforming Receivers

2.1 System model

The system model is shown in Figure 1. The details are given below:

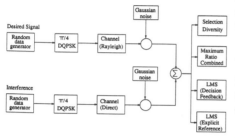

Figure 1: System Model

Receiver

We choose the selection diversity (SD) technique and the maximum ratio combined (MRC) diversity technique as the diversity receivers. The former is widely used in the actual mobile communication systems and the latter is the optimum diversity technique in Rayleigh fading environment[3]. The co-phasing of the MRC receiver is achieved by using a pilot signal [3]. We also choose the LMS beamforming algorithm as the beamforming receiver because the algorithm does not require any prior knowledge about the direction of arrival. This feature can be very useful for mobile communications.

Channel

We assume the desired signal to follow a Rayleigh distribution. One interfering signal

is considered in these simulations for the fundamental study to analyze the effect of interfering signals under actual situations. We assume the interfering signal to be the direct wave. These assumptions imply that the desired signal comes from a mobile terminal which is invisible from the base station, and the interfering signal from an adjacent base station is dominant in this situation.

Transmitted signal

The signals are modulated by using $\pi/4$-DQPSK, which is widely used for mobile communication systems. We choose the bit duration to be 16.

Antenna

The antenna array consists of 4 elements arranged in a square configuration. The inter-element spacing is assumed to be 0.5λ.

Other parameters

The simulations are run for 6250 iterations. We assume the presence of Gaussian noise in the channel. The receiver noise is not considered in these simulations. The step size of the LMS beamforming receiver is 0.01. The above simulation conditions are summarized in Table 1 below.

Table 1: Simulation Parameters

Antenna	Square array (spacing $\lambda/2$)
Receiver	SD, MRC, LMS
Modulation	$\pi/4$ DQPSK
Channel	Desired: Rayleigh Interference: Direct wave
Noise	Gaussian

2.2 Reference generation for the LMS beamformer

Generally, it is difficult to obtain a reference signal in mobile communications because the transmitted signals are not known at the receiver site. In this paper, we apply the decision feedback (DFB) to obtain a reference for the LMS beamformer. Figure 2 shows the DFB reference generation. In this method, the re-modulated signal of the decision, that is the demodulated signal, is used as the reference. This method does not require any explicit reference at the receiver site.

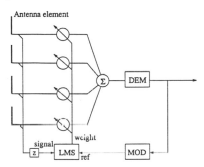

Figure 2: Reference Generating Loop

2.3 BER simulation results

Figure 3 shows the simulation results for the BER charactersitics versus signal-to-interference ratio (SIR). In this case, the direction of arrival (DOA) is 60° and the signal-to-noise ratio (SNR) is 12dB. The one element receiver and the selection diversity receiver can not achieve a BER value of 10^{-3} due to the noise floor. In this case the MRC receiver can achieve the BER value of 10^{-3} when SIR is 10dB. The LMS receiver with DFB can achieve the same value of BER when SIR is equal to 6dB. The BER of the LMS receiver with the explicit reference is less than 10^{-4} for the considered SIR.

Figure 4 shows another example when the DOA of the interfering signal is 30°. In this case, the MRC receiver can achieve better BER characteristics than the LMS receiver with DFB. The LMS receiver with explicit reference has the same characteristics as in Figure 3.

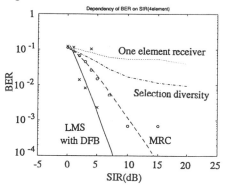

Figure 3: BER performance vs. SIR: DOA of the interference is 60°.

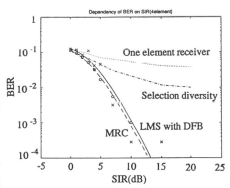

Figure 4: BER performance vs. SIR: DOA of the interference is 30°.

Next, we consider the BER characteristics when the receiver uses an 8 element circular array. The radius of the circular array is the same as the 4 element array. Figure

5 and 6 show the obtained BER characteristics when the DOA of the interfering signal is 60° and 30°, respectively. The BER performance of the LMS receiver with DFB is almost the same or better than than that of the MRC receiver. However, both are dramatically changed when SIR is higher than some value. It is because in the case of the LMS with DFB, when SIR is less than a certain value, the reference signal cannot lock to the desired signal and the weights are divergent. In the case of the MRC receiver co-phasing is achieved by using the pilot signal. So, when SIR is less than a certain value, the receiver cannot follow the pilot signal.

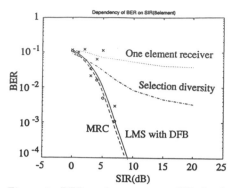

Figure 6: BER performance vs. SIR for 8 element array: DOA of the interference is 30°.

ratio combined diversity receiver.

3 Neurobeamforming for Wireless Communications

The communications literature abounds with algorithms for adaptive beamforming. The disadvantage of some of the standard methods, such as the LMS algorithm, is that they converge very slowly. In mobile communications, a fast converging algorithm is essential for adapting to the varying environments. Among the recently proposed methods, Hopfield neural network lends itself naturally for minimizing the contribution of the interfering signals at the array output. It also provides faster convergence rate than the method of LMS beamforming [4].

3.1 System model

Consider a receiving array with N antenna elements for a multi-user Code Division Multiple Access (CDMA) system. The received signal at the ith element is the sum of all the desired signals from the active users, the in-

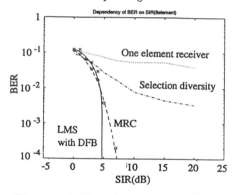

Figure 5: BER performance vs. SIR for 8 element array: DOA of the interference is 60°.

The above results indicate that the LMS beamforming receiver with the explicit reference provides a BER of less than 10^{-4} with SIR of more than 0dB and therefore, gives better performance than the diversity receivers when the desired signal is known *a priori*. When the reference signal of the LMS beamformer is generated by using decision feedback, although the BER performance depends on the situation, our simulation results show that the LMS beamforming receiver with DFB has almost the same, or better, BER performance as the maximum

terfering signals, and noise. It is given by [4]

$$S_i(t) = \sum_{k=1}^{K} D_k(t)e^{j\alpha_{i,k}} + \sum_{k=1}^{K'} I_k(t)e^{j\alpha'_{i,k}} + n_i(t),$$

$$\tag{1}$$

where K and K' are the number of active users and interfering signals. The terms $\alpha_{i,k}$ and $\alpha'_{i,k}$ represent the phase at the ith antenna relative to the reference antenna for the kth desired and interference signals, respectively. The desired signals which are assumed to be biphase modulated, and the interfering signals are of the form

$$D_k(t) = a_k e^{j(\frac{2\pi t}{T} + \phi_k(t) + \psi_k)}, \tag{2}$$

$$I_k(t) = a_k^i e^{j(\frac{2\pi t}{T} + \psi'_k)}. \tag{3}$$

Here, $\phi_k(t)$ is the modulo 2π sum of the data and code sequence, and ψ_k and ψ'_k are the initial phase for the kth desired signal and interference, respectively. A method of generating the code sequence is discussed next.

3.2 Chaotic codes

In order to spread an information signal, a completely random sequence is an ideal choice. However, this is impractical because we should be able to generate the exact sequence at the receiver end. A good choice will be to use a sequence that is random as well as deterministic. The PN sequences are commonly used for spreading signals. However, these sequences are periodic and may lead to insecure communications. A better alternative is to use chaotic sequences which exhibit good auto-cross correlation properties and are not periodic. These types of codes are particularly suited to the multi-user systems employing CDMA techniques.

We consider a chaotic sequence that is based on a logistic map of the form [4, 5]

$$F_\mu(c) = \mu c(1 - c). \tag{4}$$

Here, μ is the mapping parameter. The obtained sequence can then be converted into a symbolic form so that it is compatible with the digital communications system. In the case of bi-phase modulated signals, the sequence is partitioned into two symbols [4].

3.3 Hopfield beamformer

Now consider the Hopfield neurobeamformer shown in Figure 7. Each of the received signals at N antenna elements is split into in-phase and quadrature components. Here, we assume that only one user is present in the system. However, the same concept can be extended for a multi-user system of section 3.1.

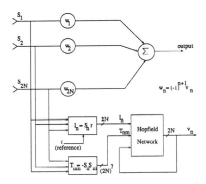

Figure 7: Hopfield Neurobeamformer

The Hopfield neural network consists of a single layer of highly connected neurons with recurrent configuration. The neurons are assumed to have input-output characteristics defined in a piecewise linear fashion. The required connectivities for the synaptic weights and external inputs of the Hopfield network are [4, 6]

$$T_{nm} = -E[S_n S_m], I_n = E[r S_n]. \tag{5}$$

Once we implement these connectivities, the mean square error between the array output

94

and the reference signal is minimized. The reference for the Hopfield neurobeamformer is generated by processing the array output appropriately, as discussed in [2].

As indicated in Figure 8, the Hopfield beamformer provides a satisfactory protection against interfering signals. The system considered in this simulation consists of an 8 element antenna array arranged in a circular configuration with an inter-element spacing of 0.7654λ. The information signal is spread by using a chaotic sequence, discussed in section 3.2, with a spreading ratio of 7:1. The desired signal is incident at $0°$ and an interfering signal is at $320°$. The obtained value of the signal-to-interference ratio is 42 dB. Thus, the neurobeamformer is able to adapt its beam pattern in response to the changing environment, thereby resulting in enhanced signal reception and higher reliability.

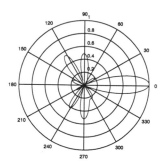

Figure 8: Linear Array: Desired signal from $0°$ and interfering signal from $320°$.

4 Conclusions

The diversity techniques require relatively simple hardware than beamforming. However, the adaptive beamforming techniques can provide higher performance gains, even in the presence of interferers. We have illustrated this point by comparing the BER for diversity techniques and the commonly used LMS beamforming. In mobile communication systems, the slow convergence of the LMS algorithm is highly undesirable. In order to overcome this problem, we have presented the Hopfield neurobeamformer that can be used in wireless communications. This combination of neurobeamforming and spread spectrum communications is expected to increase the system capacity significantly.

References

[1] K. Cho, T. Lo and J. Litva, "BER Performance Comparisions Between a Diversity Receiver and a LMS Beamforming Receiver Operating with Cochannel Interference", *WIRELESS 95 - Conference to be held in Calgary*, July, 1995.

[2] R.T. Compton, Jr., *Adaptive Antennas: Concepts and Performance*, Prentice Hall, New Jersey, 1988.

[3] W.C. Jakes, Jr., ed., *Microwave Mobile Communications*, John Wiley & Sons, New York, 1974.

[4] H. Leung, B. Quach, T. Lo and J. Litva, "A Chaotic Neural Beamformer for Wireless Communications", *PACRIM Conference*, 1995.

[5] S.N. Rasband, *Chaotic Dynamics of Nonlinear Systems*, John Wiley & Sons, 1990.

[6] A. Sandhu, T. Lo, H. Leung and J. Litva, "Hopfield Neurobeamformer for Spread Spectrum Communications", *PIMRC 95 - Conference to be held in Toronto*, Sept., 1995.

PCS System Design Issues in the Presence of Microwave OFS

Thomas T. Tran[1] , Solyman Ashrafi[2] , A. Richard Burke[3]

Moffet, Larson & Johnson, Inc.
5203 Leesburg Pike
Suite 800
Falls Church, VA 22041
Telephone: (703) 824-5660
Fax: (703) 824-5672

Abstract

There are many important technical issues concerning the design of the next generation personal communication services (PCS) systems. These issues include microcell and macrocell propagation prediction models, multiple access technologies, and interference between PCS users and microwave Operational Fixed Service (OFS) users. The purpose of this paper is to show that microcells PCS system design contributes less interference power to the OFS users than conventional system design. Robust microcell and macrocell propagation models, such as GTD and COST-231, and PCS/OFS field measurement and interference computation techniques can be used effectively to aide the PCS system design process. The propagation models use a combination of terrain, morphology, and building database for macrocell and microcell predictions and signal strength measurements for accuracy optimization. In addition, a case study is provided to exemplify the discussed supposition that microcell (base stations well below the surrounding buildings clutter with low transmitter power) system design minimizes microwave OFS interference and their immediate relocation. Practical interference calculations between different wireless systems are also discussed.

1.0 Introduction

The main issue with the PCS spectrum is that it is currently occupied by thousands of existing Operational Fixed Services (OFS) microwave facilities. The OFS community is point to point private service operators including: utility companies, railroad operators, local and federal government, police, fire and other companies and public services that do not use the spectrum for direct profit purposes. FCC legislation requires that the PCS operator protect the operation of the OFS facility from interference within the guidelines established in the TIA Bulletin 10F report. If interference cannot be avoided, then the PCS operator must negotiate with the OFS operator to relocate to either an alternative microwave band (e.g. the 6 GHz band), fiber or other telephone facility to allow for the uninterrupted communication of transmitted information. When insufficient radio channels are available for a particular site, due to an interference conflict with the microwave facility, then the PCS operator will need to remove the facility to obtain the required radio spectrum. The extent of the interference issue will be different from market to market due to the geographic distribution of OFS facilities. Additionally, the size of the PCS radio spectrum (there are three 30 MHz and three 10 MHz licenses available), the density of the traffic within the market, the modulation scheme and bandwidth of the PCS radio channel will affect the level and degree of interference into the microwave receiver.

Specialized software products have been developed that will provide analysis of interference and channel sharing coordination between PCS and OFS operations. These tools will minimize the total number of microwave facilities that need to be migrated through simulation and prediction of interference between the operations. They will also assist with the recommendation of non-interfering radio channels for the PCS station and will support the quality engineering of PCS networks.

---

[1] Project Manager

[2] Director of Advanced Technology

[3] President

The remainder of this paper will examine macrocell and microcell propagation modeling techniques for both COST-231 and the GTD (Geometrical Theory of Diffraction). These models are presented in detail along with coverage prediction and comparative drive measurements for test sites in the city of Washington, D.C. The merits and drawback of these modeling techniques are presented and analyzed in light of PCS coverage. Additionally, this paper examines prediction and measurements techniques of interference from PCS to OFS and OFS to PCS facilities and suggests the application of microcell modeling to improve interference predictions. Microcell base stations contribute much less interference power to the nearby OFS than the conventional base stations. Thus, utilizing microcells create many more available channels to be used immediately without relocating some OFS sites.

2.0 PCS and OFS Field Measurements

PCS and OFS field measurements were collected and analyzed for a small area of the Washington/Baltimore MTA. The PCS signal measurements were completed by transmitting a CW signal at 1915 MHz using a RF signal generator and a power amplifier. The receiver is made up of a K&L 4FV50 RF bandpass filter, a Sonoma Instrument 330 broadband amplifier, a HP spectrum analyzer, a GPS receiver, and a notebook PC with custom control software. For OFS signal measurements, the spectrum analyzer was programmed to scan the entire PCS spectrum with a resolution bandwidth of 1 KHz. However, a digital receiver is currently being manufactured for use at each PCS base station for semi-real time frequency sharing coordination between PCS and OFS systems.

2.1 PCS Field Measurements

Microcell and Macrocell measurements were conducted within a controlled manner. Microcells were created by extending the antenna mast from a fully equipped measurement van to a height of 12 meters, which is much less than the local average buildings height. The RF transmitter sent a CW signal at 1915 MHz using an omnidirectional antenna. The signal source came from an RF signal generator and a power amplifier followed by a bandpass filter. The mobile receiver sampled the transmitted signal continuously with a constant sampling time interval and saved the measured data. Mobile positions were obtained from the GPS and ETAK systems. The sampling rate was fast enough to sample the signal at every 1/2 wavelength at a maximum speed of 65 miles per hour. Also, a navigation system was installed in the vehicle to provide both longitude and latitude information in addition to the distance traveled, speed, and heading. The test receiver construction is rather simple: a CW RF signal is bandpass filtered, pre-amplified, and sent to the spectrum analyzer. A full size van was used for the actual drive testing with the receiver antenna height of 1.8 meters above ground.

2.2 OFS Field Measurements

In order to identify and isolate microwave signals, a directional, high gain antenna was used. The antenna was mounted on a sturdy tripod with bearing and elevation indicators to aim its main beam at the intended microwave facilities. The stationary receiver automatically made multiple scans of the entire PCS spectrum and was commanded by the notebook PC to obtain the average OFS spectrum. The performance of the spectrum analyzer alone was not suited for low level OFS signal measurements. Due to the low sensitivity of the spectrum analyzer and its broadband non-linear nature, we also had to use the SI 330 broadband amplifier and the K&L RF bandpass filter to improve performance. Precautions were made to avoid operating the amplifier in its non-linear region in case of causing excessive intermodulation and harmonic distortion. It is important to have a low noise amplifier and minimum cable or filter losses to improve the receiver sensitivity. Therefore, only high quality RF cable, RF connectors, low insertion loss filters were used, and the length of the cable was kept to minimum. The RF bandpass filter was needed to prevent strong out-of-band signal sources from reaching the spectrum analyzer mixer to keep it operating linearly. The bandpass filter's bandwidth extends from 1850 to 1990 MHz with 0.5 dB of insertion loss. The amplifier noise figure is 6 dB with 1dB compression point at 11 dBm.

3.0 Microcell/Macrocell Propagation Modeling

Propagation prediction models, using terrain and building data, are an integral part of MLJ's wireless system design tool. Buildings data can be used for both microcell and macrocell propagation predictions. This data provides much better accuracy than terrain data alone, especially for microcells (see Figure 10). However, it is equally important for accurate macrocell coverage prediction (see Figure 8).

3.1 COST-231 Microcell Propagation Model

COST-231 is a propagation model specified by the ETSI for use to design GSM systems[3]. It is composed of various microcell and a macrocell propagation modeling capabilities. The macrocell propagation prediction model is a modified version of the Okumura/Hata's model. For microcell propagation predictions, the COST-231 model uses the Walfish/Bertoni's model for multi-roof diffraction loss from the transmitter to the last roof edge before the mobile on the street. Then it uses Ikegami's model for predicting the loss from before the last roof edge down to the mobile receiver. Then, the total loss is equal to the summation of the two losses.

3.1.1 COST-231 Hata Model

This model predicts the signal strength from empirical formulas which uses different correction factors for different environments which is based on field measurements taken by Okumura in Japan[16]. Statistical formulas and correction factors of the model were derived from observation and analysis of the measured propagation data. From the measurements, Okumura generated a family of curves to predict propagation loss for various situations. Okumura also included various loss factors to account for urban losses: street orientations, terrain, mixed land and sea paths. Sometimes, it is not reliable to use these curves due to the inherent vagueness of the conditions of the correction factors. Despite their simplicity, the curves are cumbersome to use for wireless system planning. In a subsequent study, Hata was able to fit empirical formulas to Okumura's curves to efficiently incorporate them into computer programs[4]. Table 1 lists the COST-231 Hata's path loss model summary for different land types. Terrain is incorporated into the base station effective antenna height and additional diffraction losses for obstructed topography.

Table 1. COST-231 Hata's Model Summary

Frequency f:	1500 - 2000 Mhz
Base station height Hb:	30 - 200 m
Mobile height Hm:	1 - 10 m
Distance d:	1 - 20 Km
For urban and suburban area:	
L_urban(dB) = 46.3 +33.9log(f)-13.82log(Hb)-a(Hm)+[44.9-6.55log(Hb)]log(d)+Cm	
a(Hm) = [1.1log(f)-0.7]Hm-[1.56log(f)-0.8]	
Cm = 0 dB for medium sized city and suburban centers with moderate tree density	
Cm = 3 dB for metropolitan centers	
For rural quasi-open area:	
Lrqo (dB) = L_urban - 4.78[log(f)]^2+18.33log(f)-35.94	
For rural open area:	
Lro (dB) = L_urban - 4.78[log(f)]^2+18.33log(f)-40.94	

3.1.2 COST-231 Walfish/Ikegami Model

The Walfish's model is based on physical optics solution of the forward multiple diffraction past many equally spaced absorbing half-screens with equal height and uniform separation distance much larger than a wavelength[21]. Therefore, the absorbing screens diffraction can be used to model building diffraction in urban environment. Basically, the incident field on successive edges is derived using physical optics to yield a multidimensional Fresnel integral, which is expanded into a series of Boersma functions. Thus, the field solution can be obtained recursively. The Ikegami's model uses single roof diffraction and diffraction-reflection modes to calculate the mean signal strength at the receiver between two rows of buildings[5]. The first received electric field component is only diffracted by the last building before the receiver. The second field component has undergone diffraction by the last building before the receiver and reflection by the next building after the receiver. The diffraction wedge of the roof before the receiver is treated as a knife edge and building reflection is modeled using geometrical optics. The summary of this model is shown in Table 2[3].

Table 2. COST-231 Walfish/Ikegami's Model Summary

Frequency f:	800 - 2000 Mhz
Base station height Hb:	4 - 50 m
Mobile height Hm:	1 - 3 m
Distance d:	0.02 - 5 Km
Height of buildings Hroof (m)	
Width of road w (m)	
Building separation b (m)	
Road orientation with respect to the direct radio path Phi	
Non-line of sight condition between base and mobile:	
Lb = Lo+Lrts+Lmsd(or Lb = Lo for Lrts+Lmsd <= 0 dB)	
Lo is the free space path loss:	
Lo = 32.4+20log(d)+20log(f)	
***Lrts: Roof to street diffraction and scatter loss* (Ikegami)**	
Lrts = -16.9-10log(w)+10log(f)+20log(Hr-Hm)+Lcri	
Lcri = -10+0.354phi	for 0 <= phi < 35 degrees
Lcri = 2.5+0.075(phi-35)	for 35 <= phi < 55 degrees
Lcri = 4-0.114(phi-55)	for 55 <= phi < 90 degrees
***Lmsd: Multiscreen diffraction loss* (Walfish)**	
Lmsd = Lbsh+ka+kd*log(d)+kf*log(f)-9log(b)	
Lbsh = -18log(1+Hb-Hroof)	for Hb > Hroof
Lbsh = 0	for Hb <= Hroof
ka = 54	for Hb > Hroof
ka = 54-0.8(Hb-Hroof)	for d >= 0.5 and Hb <= Hroof
ka = 54-0.8(Hb-Hroof)(d/0.5)	for d < 0.5 and Hb <= Hroof
kd = 18	for Hb > Hroof
kd = 18-15(Hb-Hroof)/Hroof	for Hb <= Hroof
kf = -4+0.7(f/925 - 1)	for medium sized cities and suburban centers with moderate tree density
kf = -4+1.5(f/925 - 1)	for metropolitan centers
Line of sight condition between base and mobile:	
Lb = 42.6+26log(d)+20log(f)	for d >= 0.02 Km

3.2 GTD Propagation Model

The GTD propagation model is basically a combination of ray optics and geometrical theory of diffraction. The geometrical theory of diffraction is a heuristic extension of geometrical optics for diffraction loss prediction. The model is composed of many rays such as direct, reflected, diffracted, and scattered as well as the combination of different rays. It uses a combination of terrain, building, land-use-land-cover, and actual propagation measurement data to optimize the model prediction accuracy. The main advantage of this model is its location specific accuracy since the model basically simulates the actual RF wave propagation mechanics. Figures 8 and 10 present sample prediction analysis for the GTD microcell models incorporating a digitized building database.

3.2.1 Direct Ray

The direct ray is the ray that propagates directly from the source to the receiver without any obstruction along its path. The field strength of this ray at the receiver can be accurately approximated by the Friis free space formula which is the simplest of all propagation modes. Assuming both the transmitting and receiving antennas are isotropic, then the received power can be computed by,

$$P_{Rx-freespace} = (\lambda / 4\pi d)^2 \times P_{Tx} \quad [1]$$

where d is the T-R separation distance. Therefore, the free space path loss formula in dB is

$$PL_{dB-freespace} = 20 \times \log_{10}(\lambda / 4\pi d).$$

3.2.2 Specular Reflection

The electric field at the receiver is simulated by tracing specular reflected rays from the transmitter to the receiver. The total received power is the summation of the power of each individual ray. High frequency approximation of reflected fields is based on Snell's law, where the specular reflected ray angle is the same as the incident ray angle. The reflection surface is assumed to have the electrical characteristics of a infinite dielectric slab with random roughness. The reflected electric field is a function of the reflection coefficient of the dielectric slab, which depends on the incident angle, material permittivity and conductivity, polarization, and surface roughness. In urban environments, building reflection coefficients vary randomly due to a variety of lossy dielectrics and roughness of building surfaces. The reflected electric field at the reflecting surface can be described by

$$\left|\bar{E}\right| = \frac{r_0}{r}\left|\Gamma\right|\left|\bar{E}_o\right|. \quad [2]$$

3.2.3 Diffraction Rays

Diffraction over a ridge, a mountain, or a building can be modeled using the geometrical theory of diffraction (GTD). GTD differs from the knife edge diffraction by accounting for the shape and electrical characteristics of the diffracting edge such as conductivity, permittivity, and surface roughness. GTD also accommodates the effect of polarization of the propagating wave, multiple diffractions, and reflections from the wedge surface. The original GTD solution formulated by Keller does not work well in the transition zone, i.e. when the source, the diffracting edge, and the receiver lie on a straight line[8]. Later, Kouyoumjian and Pathak modified Keller's work with a new diffraction coefficient formulation that also works in the transition zone[9]. GTD is an attractive model since it blends well with geometrical optics in which the diffraction coefficient plays a similar role as the reflection coefficient. However, the computation of the GTD diffraction algorithm is much more complicated than the knife edge diffraction algorithm. In general, the GTD model is better than the knife edge model for large wedge angles and in deeply shadowed locations.

The GTD[13] model predicts the electric field at a receiving point P by

$$E_{GTD} = E_o \frac{e^{-jkr}}{s} \prod_i A_i \times D_i \quad [3]$$

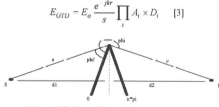

Figure 1. Diffraction wedge geometry

where A_i is the amplitude factor of the expanding wave front and D_i is the GTD diffraction coefficient for the ith wedge (also a function of polarization). Also, E_o is the source electric field amplitude at d_o, s is the distance from the source to the first diffracting wedge tip as shown in Figure 1, and d_{tot} is the total ray distance from source to receiving point ($d_{tot} = s+p$ in Figure 1).

For the single diffraction case, the amplitude factor becomes

$$A_O = \sqrt{\frac{s}{p(s+p)}}. \quad [4]$$

Also, the GTD diffraction coefficient D_i is a function of polarization of the incident field at the wedge. The original GTD development was formulated assuming the diffraction wedge is a perfect conductor, and the receiver is not close to the transition zone, then D_i for the ith wedge is given by

$$D_i^{\perp} = \frac{e^{-j\frac{\pi}{4}\sin(\pi/n)}}{n\sqrt{2\pi k}} \left\{ \frac{1}{\cos(\pi/n) - \cos((\phi-\phi')/n)} \mp \frac{1}{\cos(\pi/n) - \cos((\phi+\phi')/n)} \right\} \quad [5]$$

where all parameters are also shown in Figure 1. For the finite conducting wedge with locally rough wedge, the diffraction coefficient D_i is

$$D_i^{\perp} = \frac{-e^{j\pi/4}}{2n\sqrt{2\pi k}} \times \left[\begin{array}{l} \cot(\frac{\pi+(\phi-\phi')}{2n})F(kL_ia^+(\phi-\phi')) + \cot(\frac{\pi-(\phi-\phi')}{2n})F(kL_ia^-(\phi-\phi')) + \\ R_o^{\perp} \cot(\frac{\pi-(\phi+\phi')}{2n})F(kL_ia^-(\phi+\phi')) + R_n^{\perp} \cot(\frac{\pi+(\phi+\phi')}{2n})F(kL_ia^+(\phi+\phi')) \end{array} \right] \quad [6]$$

where $F(x)$ is the Fresnel integral,

$$F(x) = 2j\sqrt{x}e^{jx} \int_{\sqrt{x}}^{\infty} e^{-jt^2} dt \quad [7]$$

and $L_o = \frac{sp}{s+p}$ [8] for the single diffraction case.

$$\text{Also, } a^{-+}(\beta) = 2\cos\left[\frac{2n\pi N^{-+} - \beta}{2}\right]^2 \text{ and } \beta = \phi \pm \phi'. \quad [9]$$

The N-+ variables of the above equation must satisfy the conditions

$$2\pi n N^+ - \beta = \pi \text{ and } 2\pi n N^- = -\pi.$$

R_o and R_n are the reflection coefficients for perpendicular and parallel polarization of the 0th face and the nth face, respectively. These reflection coefficients are dependent on the surface conductivity and roughness. If the surface is perfectly smooth and perfectly conducting, then the reflection coefficients are +- 1 for R_o and R_n, respectively.

It is interesting to note that for n=2 and $d_1, d_2 > u$ (refer to Figure 1), the GTD solution approaches that of the knife edge solution despite their different formulations. The basic knife edge approach integrates over the fields at the diffracting aperture to yield the diffracted field. The GTD approach is the combination of ray propagation via the direct path, reflected path, and diffracted path. GTD predicts more attenuation for perpendicular and less attenuation for parallel polarization than knife edge diffraction assuming a perfectly smooth conducting wedge. However, actual terrain diffraction measurements show little difference between two polarization. If the wedge is no longer assumed to be perfectly smooth and conducting, then the GTD predictions also yield similar results for both polarizations. The GTD model is also capable of multiple diffraction prediction[12]. Figure 2 shows a typical situation of a double wedge diffraction geometry. Basically, the diffraction coefficients are found by treating the second diffracting wedge top as the receiving point for the first wedge diffraction. Then the first diffracting wedge tip as the source and the third diffracting wedge tip as the receiving point for the second wedge diffraction, and so on.

Figure 2. Multiple diffraction wedge geometry

The basic GTD equations still remain the same except for A_n and L of Equations 4 and 8, respectively. Refer to Figure 2, there are two diffracting wedges for which two diffraction coefficients must be computed. The first diffraction coefficient can be calculated by considering the ray from source S diffracted by wedge #1 to the top of wedge #2, denoted as AS12 & LS12,

$$A_{S12} = \sqrt{\frac{s}{s'(s+s')}} \; and \; L_{S12} = \frac{ss'}{s+s'} . \quad [10]$$

The second diffraction coefficient can be found by considering the ray from the source at the wedge tip #1 diffracted by wedge #2 to P, denoted as A12P & L12P,

$$A_{12P} = \sqrt{\frac{s+s'}{p(s+s'+p)}} \; and \; L_{12P} = \frac{ps'}{p+s'} \quad [11]$$

and the final diffraction coefficient is just simply the product of all computed diffraction coefficients.

3.2.4 Scattered Rays

For OBS topographies, the signal scattering and penetration also play a significant role in the total received signal strength especially when the receiver is deeply shadowed. Therefore, the penetration of signals between obstructing obstacles is considered as well as the signal scattering. The penetration loss or scattering coefficient can be determined empirically from measurements at different frequencies and from the published literature for similar studies.

3.2.5 Total Received Power

Finally, all intercepted rays at the receiver are summed in power to produce the local average received power,

$$P_{Rx-total} = P_{direct} + \sum_n P_{refl_i} + \sum_m P_{diffr_j} + P_{scatter} . \quad [12]$$

3.3 Predictions versus Measurement Results

As can be seen from Figure 7, 8, 9, and 10, the GTD model's predictions compare favorably with measurements for both microcell and macrocell studies with the standard deviation of prediction error of 7.2 dB. The waveguiding propagation effects can be observed for the microcell case. Figure 11 illustrates the COST-231 Walfish/Ikegami microcell model coverage prediction which usually under predicts the signal strengths in most area due to the absence of scattering mode. However, this problem can be remedied by optimizing the model's coefficients with measurements. Also, spatial database errors can greatly influence the outcome predictions in some cases. The COST-231 model, while not as robust as GO/GTD, is much easier to implement and use with only limited building data information needed. The four most important parameters for COST-231 Walfish/Ikegami's model are the road width around the receiver, the average roof height of buildings between the Tx-Rx path, the average building separation distance, and the building orientation with respect to the Tx-Rx path. This can be entered by the user or extracted from the available building database. On the other hand, the GO/GTD model requires the digitized building data before any prediction can be performed.

4.0 Interference Studies Between OFS and PCS

In order to minimize degradation of the existing OFS system performance, it is desirable to monitor, record, and analyze the microwave OFS signal strengths from all PCS sites to improve the interference avoidance process. The measurements also allow characterization of the microwave OFS channel usage to minimize interference to the OFS facilities and improve the analysis of PCS system performance under severe interference conditions.

4.1 Interference Computations Methodology

In order to have successful intersystem frequency coordination, interference power level (IPL) from PCS to OFS and from OFS to PCS must be measured or predicted accurately. The basic problem arises when one system is narrowband and the other system is wideband, complicated with different modulation standards and interference tolerance capabilities. For PCS systems, the modulation scheme will be digital, usually M-ary Phase Shift Key (MPSK). For OFS systems, the modulation schemes are both FM analog and digital. Given the wide variety of different power spectral density (PSD) for different modulation formats coupled with different bandwidths and filter characteristics, only the average of the IPL was computed numerically.

4.1.1 IPL Computations

For the IPL calculations, let's assume the interfering signal PSD and the receiver filter response characteristics are known. Figure 3 shows the received interfering signal power spectral density function, PSD(f) and the magnitude (power) response function of the receiver filter, H(f). Note that PSD(f) represents the received interfering signal PSD from either the OFS or PCS systems and H(f) represents either the OFS or the PCS filter response. Then, the following equations can be applied to compute the average IPL from PCS base stations to OFS sites,

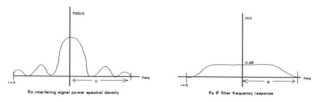

Figure 3. Typical example of Rx interfering signal PSD and IF filter frequency response

$$IPL = \int_\infty PSD(f - f_{c_{pcs}})H(f - f_{c_{pofs}})df \text{, but can be simplified to}$$

$$IPL = \int_{f_{c_{pcs}} - \frac{BW_{pcs}}{2}}^{f_{c_{pcs}} + \frac{BW_{pcs}}{2}} PSD(f - fc_{pcs})H(f - fc_{pofs})df . \quad [13]$$

The BW_{pcs} is the RF channel bandwidth of the PCS system, and f_c is the center frequency of the RF channel. Note that the above equations assume continuous PSD function per Hz which is impractical for numerical computation. Therefore, a discrete frequency (PSD per Δf (Hz)) version of the above equation for OFS to PCS interference can be used,

$$IPL = \sum_{i=0}^{N-1} H_{pcs}(i)PSD_{pofs}(i + \frac{1}{\Delta f}\left\{fc_{pcs} - B_{pcs} - fc_{pofs} + C_{pofs}\right\}) \quad [14]$$

where N is the total number of Δf steps from fc_{pcs}-B/2 to fc_{pcs}+B/2, B is half of the PCS channel bandwidth, and C is half of the OFS bandwidth. The above equation can also be used for IPL calculation from PCS to OFS interference,

$$IPL = \sum_{i=0}^{N-1} PSD_{pcs}(i)H_{pofs}(i + \frac{1}{\Delta f}\left\{fc_{pcs} - B_{pcs} - fc_{pofs} + C_{pofs}\right\}) \quad [15]$$

where B is now half of the PCS bandwidth and C is now half of the OFS filter bandwidth. Note that it is only necessary to integrate or sum over the PCS system bandwidth since the PCS Tx/Rx filters have a very fast roll-off rate, thus minimizing computational time. As an example of the typical IPL calculation in a communication link using the above methodology, consider the conventional IPL calculation approach where the PSD of the narrowband Tx signal is approximated as a CW carrier. Also assuming isotropic Tx/Rx antennas, no polarization and cable losses, then the IPL from a CW signal can be approximated as,

$$IPL = EIRP + H_{Rx} - PL \quad [16]$$

where PL is the link path loss and H_{Rx} is the receiver filter isolation as a function of frequency separation. Thus,

$$IPL = \left[\sum_{i=0}^{N-1} PSD_{Tx_{pcs}}(i) H_{pofs}\left(i + \frac{1}{\Delta f}\left\{ fc_{pcs} - B_{pcs} - fc_{pofs} + C_{pofs}\right\}\right)\right] - PL \quad [17]$$

where PSD_{Tx} is the transmitted signal PSD at the PCS transmitter.

The post processing of the measured data reconstructs the measured OFS PSD. For example, a OFS spectrum of 5 MHz can be reconstructed using 200 25KHz channels. This technique allows for detail study of the OFS spectrum as well as the OFS filter characteristics which can interfere with PCS operations. Also, since there are many captures of the same spectrum over time, a rough statistical spectral study of OFS transmitting activities over time is feasible. The power content of a wideband signal within a frequency span of $N\Delta f$ from measurements is

$$\overline{Pt}_{span_{dBm}} = 10\log_{10}\left(\sum_{i=0}^{N-1} 10^{.1\overline{Pt}_{i_{dBm}}}\right). \quad [18]$$

Also,

$$P_{Rx_{mW}} = P_{Tx}G_{Tx}(q_{Tx},f_{Tx})G_{Rx}(q_{Rx},f_{Rx})(\frac{\lambda}{4\pi d})^2 (AddAtten) \quad [19]$$

and the path loss for a given communication link is

$$PL_{dB} = 20\log_{10}(\frac{\lambda}{4\pi d}) + AddAtten_{dB} \quad [20]$$

where *AddAtten* is the additional path loss or enhancement due to obstructions or reflections, and G_{Tx} is the OFS antenna gain in the direction of the PCS site. Thus, the OFS-PCS path loss can be computed by

$$PL_{dB} = \overline{Pt}_{Span_{dBm}} - P_{Tx_{dBm}} - G_{Tx}(q_{Tx},f_{Tx})_{dB} - G_{Rx}(q_{Rx},f_{Rx})_{dB}. \quad [21]$$

where P_{Tx} is the transmitting OFS power. Using the principle of propagation reciprocity, the PCS to OFS path loss should be identical. One should be careful to note that the actual path loss of the PCS to OFS link is not exactly the same as the OFS to PCS link since they are not operating at the same frequency and the channel is naturally time variant. However, this assumption is valid since the separation of the Tx/Rx frequencies are much smaller than the actual carrier frequency. Thus, the analysis of measurement data can be used to compute the actual interference from OFS to PCS and from PCS to OFS to optimize the process of spectrum resources allocation.

4.1.2 PSD and Filter Responses

The receiver filter frequency response functions can be obtained from the radio equipment manufacturers. The average PSD of the interfering signal can be obtained in three ways: radio system manufacturers, compute/simulate, or actual field measurements. However, computation of the interfering signal PSD is not necessarily accurate and the transmitter RF filter characteristics are not generally known. A better approach is to measure the PSD of a known OFS transmitting source and characterize it according to the modulation scheme and the allocated bandwidth. This approach will likely yield more realistic average PSD estimates than the computational method. However, the measured signal should have at least 60 dB of dynamic range to be effective. Thus, the OFS monitoring receiver sensitivity must also be very good.

4.1.3 Interference Study from PCS to OFS

There are two microwave spectrum bands around 2 GHz frequency to be allocated to licensed broadband PCS system providers by the FCC. The lower spectrum band is used for the PCS subscriber unit uplink and the upper band is used for the base station down link as shown in Figure 4,

Figure. 4 Microwave spectrum utilization

where the OFS link can transmit and receive in both PCS bands. A typical OFS and PCS coexistence scenario is illustrated in Figure 5.

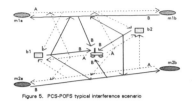

Figure 5. PCS-POFS typical interference scenario

Figure 5 illustrates an important scenario when developing approaches/algorithms for interference analysis. There are two PCS base stations (squares) and two OFS sites (ellipses) with a total of 4 independent links (2 OFSs and 2 PCSs). At each receiver, there are two potential intersystem interferers which must be studied to see if they can coexist. The interference studies can be performed by direct measurements and/or propagation predictions. Two categories of measurements are required for successful PCS system design to minimize intersystem interference: measurements at PCS base stations and at the mobiles are used to compute the PCS->OFS and OFS->PCS interference. The OFS->PCS interference can be obtained in real time. This would also be true if a radio monitoring unit (RMU) was placed at each OFS site to measure PCS signal strengths in the downlink B band. Since the RMU is only designed to operate in the A band (uplink), path losses can be obtained for links (b1-m1a), (b2-m1a), (b1-m2b), and (b2-m2b) as shown in Figure 5. Path losses for links (b1-m1b), (b2-m1b), (b1-m2a), and (b2-m2a) are not needed since OFS sites m1b and m2a do not interfere with PCS base stations but rather with PCS mobiles. However, these OFS links are subjected to interference from the mobiles which is harder to quantify since the number of mobiles, their locations, and their distributions are random. Our immediate efforts will concentrate on interference studies between the OFS sites and the PCS base stations.

In this example, the RMU will be placed at b1 and b2 base stations to measure the signal strengths from both the m1a and m2b facilities. Figure 6 shows the typical measurement signal characteristics.

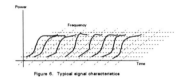

Figure 6. Typical signal characteristics

The RMU will be able to record snapshots of the signal power in the assigned frequency slots over some period of time. Interference from PCS to OFS can be calculated using the theory of path loss reciprocity where the path loss is the same for the forward and reverse link at any instant of time operating at the same frequency. The total PCS interfering power at any OFS site is the sum of the contributing power from all the operating PCS base stations. However, the total PCS interfering power must not exceed a certain threshold (cut-off) at any OFS sites, which ultimately determines the availability of spectrum for the start-up PCS system.

4.2 PCS Channel Availability Study

The total amount of interference at a particular OFS site is the sum of all the received interference power from all PCS users. Therefore, PCS channel planning and reuse has to incorporate the current OFS users to avoid both intersystem and intra-system interference. PCS channel availability also depends on multiple access technologies such as TDMA or CDMA. In order to determine the availability of a particular channel for either CDMA or GSM systems, that channel is activated at each requested base station to see if its radiating power exceeds the interference limit at any OFS facilities. In particular, the CDMA standard allows the same channel to be assigned to as many locations as possible. However, the GSM standard, channels are assigned based on an optimum frequency reuse pattern to minimize the interference. The channels are assigned through out the entire system starting with dense urban base stations, then moving to urban base stations, and so on through suburban and rural base stations. In the first pass, the highest ranked available channel is assigned to all sectors. Then, all channels are assigned recursively until either no more channels are required by the PCS base station or no more channels can be assigned without crossing the cut-off limit of any OFS.

In order to optimize frequency coordination between OFS and PCS users, OFS measurement data taken at PCS base stations can be used to adjust the predicted interference levels between the two systems. Channel availability can be more accurately obtained from the measurements than from predictions. Thus, more channels can be confidently assigned to the PCS sites without crossing the IPL cut-off thresholds. In some cases, enough channels could be found using measurement data to delay the relocation of some OFS sites. The delay in the relocation of the OFS sites reduces PCS start-up cost which will make PCS more competitive with the current cellular system.

5.0 Case Study

A system design case study was conducted at the Washington Post building in downtown Washington DC. The study was performed to see the interference effects from a macrocell and a microcell PCS base station to the nearby OFS facilities. Table 3 lists the OFS site properties and Table 4 lists the PCS site properties. Although the interference effect is a function of modulation and bandwidth, the average interference power cut-off level can be used as a measure of the radio system interference tolerance. The cut-off IPL is dependent on the type of radio receiver and manufacturer. The IPL from the PCS base stations to the OFS sites and channel availability were computed and PCS channel assignments are obtained such that the OFS IPL is never exceeded. Note that the IPL is a function of EIRP, path loss, filter, antenna, and transmission line, which is obtained from the summation of all cochannel and adjacent channel interference power from all PCS base stations. Figure 12 illustrates the map of nearby OFS links and prospective PCS sites for the study. Also, Figure 13 shows the macrocell OFS spectrum measurements which can be used to improve the interference calculations accuracy.

Table 3. OFS site properties to be studied

MW Site	Rx Call	Tx Call	Frequency	CutOff	Status	Coordinates
34995	WNEN376	WNEN377	1960 MHz	-83.5 dBm	ON	38 53 39.0N and 77 4 11.0W
40809	WNTL678	WNTM952	1960 MHz	-91.3 dBm	ON	38 53 26.0N and 76 52 10.0W

Table 4. Candidate PCS base station properties

PCS Site	Coordinates	EIRP	Antenna Height
1B (Wpost street-microcell)	38 54 12.0N 77 2 10.0W	24 dBm	11.9 meters
3B (Wpost roof-macrocell)	38 54 13.0N 77 2 9.0W	50 dBm	40 meters

Table 5. PCS-OFS intersystem interference study report

Interference Report #1		Interference Report #2	
PCS Site ID : 1B	Rx MW Site ID : 34995	PCS Site ID : 3B	Rx MW Site ID : 34995
Sector ID : 1	Rx Call Sign : WNEN376	Sector ID : 1	Rx Call Sign : WNEN376
PCS Coordinates : 38 54 12.0N	Rx MW Coordinates : 38 53 39.0N	PCS Coordinates : 38 54 13.0N	Rx MW Coordinates : 38 53 39.0N
: 77 2 10.0W	: 77 4 11.0W	: 77 2 9.0W	: 77 4 11.0 W
PCS Power (dBm) : 24	Dist to Tx MW (km) : 20.3	PCS Power (dBm) : 50	Dist to Tx MW (km) : 20.3
Antenna Height (m) : 11.9	Az to Tx MW (deg) : 26.7	Antenna Height (m) : 40.0	Az to Tx MW (deg) : 26.7
Interference Type : Base station	Az to PCS (deg) : 70.7	Interference Type : Base station	Az to PCS (deg) : 70.3
Dist to Rx MW (km) : 3.1	Frequency (MHz) : 1960.0	Dist to Rx MW (km) : 3.1	Frequency (MHz) : 1960.0
Site Type : Dense urban	Cutoff (dBm) : -83.5	Site Type : Dense urban	Cutoff (dBm) : -83.5

Az to Rx MW (deg) : 250.7	IPL from PCS (dBm) : -83.0	Az to Rx MW (deg) : 250.3	IPL from PCS (dBm) : -57.0
Interfering Channels : 1	Margin (dB) : -0.5	Interfering Channels : 1 2 3 4 5	Margin (dB) : -26.5
Channels that are on : 14 18 24 25	Rx MW Status : On	: 6 7 8 9 10	Rx MW Status : On
Channels that are off : 2 3 4 5 6	Tx Call Sign : WNEN377	: 11 12 13 14 15	Tx Call Sign : WNEN377
: 7 8 9 10 11	Tx MW Site ID : 34996	: 16 17	Tx MW Site ID : 34996
: 12 13 15 16 17	Tx MW Coordinates : 39 3 27.0 N	Channels that are on : 25	Tx MW Coordinates : 39 3 27.0N
: 19 20 21 22 23	: 76 57 50.0W	Channels that are off : None	: 76 57 50.0W
Channels Not Avail : None		Channels Not Avail : None	

From the OFS database search, there is one major interfering OFS path near the PCS base station: WNEN376/WNEN377. The OFS sites and the PCS sites were setup to share a common frequency band (cochannel). The study was done specifically for GSM system where 200 KHz RF channels are requested and the study results are shown in Table 5. From the interference study, this OFS path would incur severe interference from PCS site 3B, the base station on the roof of the Washington Post building. As can be seen from the detailed study Report #2, there is only one available channel out of 25 requested channels. However, the microcell base station in the same area contributed about 26dB less IPL to the same OFS than the macrocell base station. Report #1 shows that there are four immediate available channels, twenty channels are possibilities, and only one channel is permanently excluded. Clearly, the macrocell 3B cannot coexist with the current OFS link WNEN376/WNEN377 unless its EIRP is dramatically reduced and/or its antenna height is lowered. By reducing the PCS cell coverage radius using buildings as natural electromagnetic field (EMF) shields, relocation of this OFS link may not be necessary due to the reduced IPL to the OFSs.

6.0 Conclusion

There are several technical issues to be resolved before two industries can share the same spectrum. The primary problem is protecting the quality operation of the incumbent microwave facilities during the three-five year period of spectrum sharing. This interference avoidance has been a major focus of the Telecommunications Industry Association (TIA), which develops technical standards for the telecommunications industry. If the emerging PCS operator cannot protect the operation of the OFS microwave facilities, then the PCS operator must either find an alternative radio channel or relocate the OFS facilities to an alternative medium.

In designing PCS systems, field measurements should also include an OFS signal strength spectrum scan at each base station. This step is necessary to optimize the frequency coordination between the existing OFS facilities and PCS base stations. Accurate propagation models play an important role in the ultimate success of PCS deployment which shares the same spectrum with the existing microwave OFS. Microcell propagation models, using digitized building data, should be used for microcell base station design and interference studies to maximize efficient use of the scarce spectral resources and to optimize system performance. Such models include GTD and COST-231 which are robust and can be used effectively for both macrocell and microcell conditions. The average IPL which determines the interference tolerance between different communication systems were computed using the interfering signal power spectral density and the receiver filter response. From a specific case study of cochannel intersystem interference, microcell base stations can be used effectively to locate more usable spectrum which minimizes the relocation of some OFS facilities.

Acknowledgment

The authors wish to acknowledge Mary Compton for her editing suggestions and Juan Portillo for his efforts in running the interference studies and plotting the coverage maps.

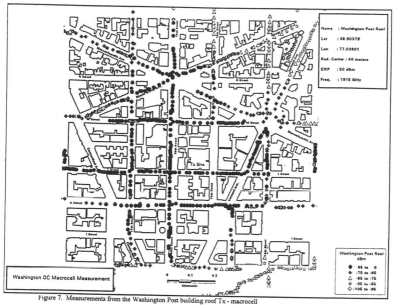

Figure 7. Measurements from the Washington Post building roof Tx - macrocell

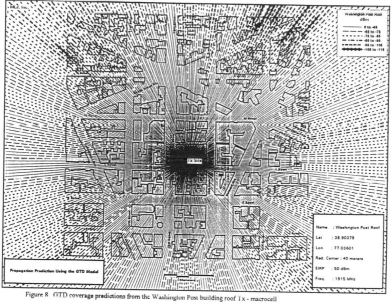

Figure 8. GTD coverage predictions from the Washington Post building roof Tx - macrocell

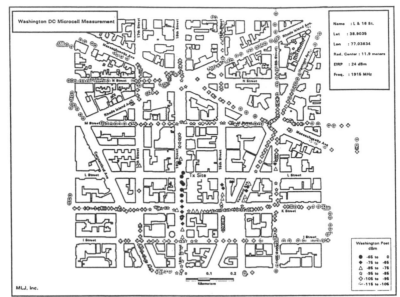

Figure 9. Measurements from the Washington Post building street Tx - microcell

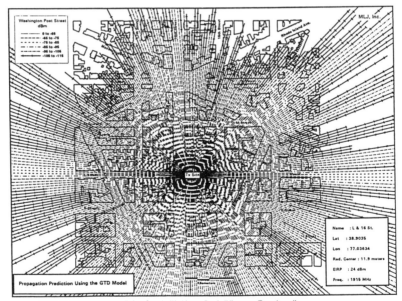

Figure 10. GTD coverage predictions from the Washington Post building street Tx - microcell

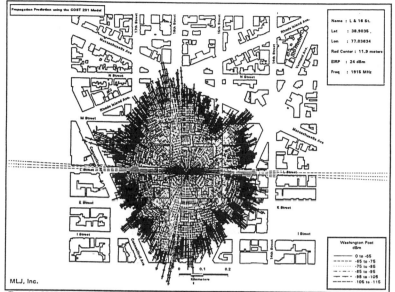

Figure 11. COST-231 Walfish/Ikegami coverage predictions from the Washington Post building street Tx - microcell

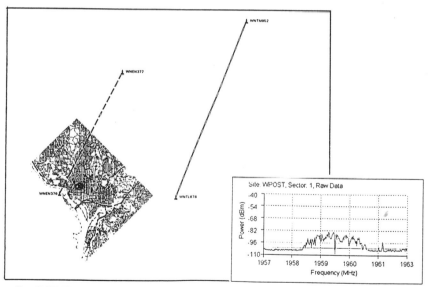

Figure 12. PCS-OFS site map in the Washington DC area

Figure 13. OFS spectrum measurement

REFERENCES

[1] B. Bisceglia, "Symbolic Code Approach to GTD Ray Tracing" IEEE Transactions on Antennas and Propagation, Vol. 36, No. 10, pp. 1492-1495, Oct. 1988.

[2] K.A. Chamberlin, et al,"An evaluation of Longley-Rice and GTD propagation models",IEEE trans. on ant. & prop., Nov 1982.

[3] ETSI-SMG Technical Specification, "European digital cellular telecommunication system (phase 2:Radio network planning aspects", ETSI-SMG GSM 03.30 version 4.1.0, January 1993.

[4] M. Hata, "Emperical Formula for Propagation Loss in Land Mobile Radio Services," IEEE Transactions on Vehicular Technology, Vol. VT-29, No. 3, pp. 213-225, August 1979.

[5] F. Ikegami, et. al., "Propagation factors controlling mean field strength on urban streets," IEEE Transactions on Antennas and Propagation, Vol. AP-32, No. 8, pp. 822-829, Aug. 1984.

[6] F. Ikegami, T. Takeuchi, and S. Yoshida., "Theoretical Prediction of Mean Field Strength for Urban Mobile Radio," IEEE Transactions on Antennas and Propagation, Vol. 39, pp. 299-302, Mar. 1991.

[7] J.B. Keller, "Diffraction by an Aperture," Journal of Applied Physics, vol. 28, April 1957, pp. 426-444.

[8] J.B. Keller, "Geometrical Theory of Diffraction," Journal of the Optical Society of America, Vol. 52, No. 2, Feb. 1962, pp. 116-130

[9] R.G. Kouyoumjian and P.H. Pathak, "A Uniform Geometrical Theory of Diffraction for an Edge in a Perfectly Conducting Surface," Proceedings of the IEEE, vol. 62, Nov. 1974, pp. 1448-1461.

[10] M. C. Lawton and J. P. McGeehan, "The application of GTD and Ray Launching Techniques to Channel Modelling for Cordless Radio Systems," 42nd IEEE Veh. Tech. Conf., Denver, May 1992, pp. 125-130.

[11] A. G. Longley and P. L. Rice, "Prediction of Tropospheric Transmission Loss Over Irregular Terrain - A Computer Method," ESSA Technical Report, ERL 79IT567, 1968.

[12] R.J. Luebbers, "Finite Conductivity Uniform GTD Versus Knife Edge Diffraction in Prediction of Propagation Path Loss," IEEE Transactions on Antennas and Propagation, Vol. AP-32, pp. 70-76, Jan. 1984.

[13] R.J. Luebbers, "Propagation Prediction for Hilly Terrain Using GTD Wedge Diffraction," IEEE Transactions on Antennas and Propagation, Vol. AP-32, pp. 951-955, Sept. 1984.

[14] R.J. Luebbers, and et al, "Comparison of GTD Propagation Model Wide-BandPath Loss Simulation with Measurements," IEEE Transactions on Antennas and Propagation, Vol. 37, pp. 499-505, Apr. 1989.

[15] A. R. Noerpel, "Use of Physical Optics to Characterize Building Reflections," in proceedings for IEEE Int. Conf. Commun., (ICC'86), pp. 847-851, 1986.

[16] Y. Okumura, et. al., "Field Strength and its Variability in VHF and UHF Land Mobile Service," Rev. Elec. Comm. Lab., Vol. 16, pp. 825-873, Sept.-Oct.1968.

[17] A. Ranade, "Local Access Radio Interference Due to Building Reflections," IEEE Transactions on Communications, Vol. 37, No. 1, pp. 70-74, January 1989.

[18] T.S. Rappaport, "Characterization of UHF Multipath Radio Channels in Factory Buildings," IEEE Transactions on Antennas and Propagation, Vol. 37, No. 8, August 1989, pp. 1058-1069.

[19] S.Y. Seidel, et.al, "Path Loss, Scattering, and Multipath Delay Statistics in Four European cities for Digital Cellular and Microcellular Radiotelephone," IEEE Transactions on Vehicular Technology, Vol. 40, No. 4, Nov. 1991, pp. 721-730.

[20] S.Y. Seidel, et.al, "The Impact of Surrounding Buildings on Propagation for Wireless In-Building Personal Communications System Design", 42nd IEEE Vehicular Technology Society Conference, Denver, CO, May 1992.

[21] J. Walfish and H. L. Bertoni, "A Theoretical Model of UHF Propagation in Urban Environments," IEEE Transactions on Antennas and Propagation, Vol. 36, pp. 1788-1796, Dec. 1988.

[22] H.H. Xia and H. L. Bertoni, "Diffraction of Cylindrical and Plane Waves by an Array of Absorbing Half-Screens," IEEE Transactions on Antennas and Propagation, Vol. 40, pp. 170-177, Feb. 1992.

An Evaluation Point Culling Algorithm
for Radio Propagation Simulation based on the Imaging Method

Satoshi TAKAHASHI *, Kazuhito ISHIDA *,
Hiroshi YOSHIURA ** *and* Arata NAKAGOSHI *

* Hitachi, Ltd. Telecommunications Division
** Hitachi, Ltd. Systems Development Laboratory

Abstract *This paper presents a three-dimensional radio propagation simulation technique that reduces its computation time. The technique is based on the imaging method, which is one of the ray tracing methods. First, the probability that propagation paths exist at evaluation points within second-order reflection is shown. The result shows that the probability is less than a few percents. Therefore, it appears that the computation time can be reduced by eliminating the calculation for evaluation points where rays never launch. Second, the algorithm implemented the idea is presented. The algorithm uses "approximated illumination area." There aren't any evaluation points outside the area. The area is solved from the coordinate relationships between the last reflected object and the last imaging point. The last imaging point is derived from recursive image generation of the transmitting point. After that, only evaluation points within the illuminated area are calculated. The computation time can be reduced to less than one-quarter of the time previously required because the calculation for evaluation points where rays don't launch never be executed. The accuracy of the proposed method is not degraded since the method just eliminates redundancy of the path determination part of the conventional imaging method.*

1. Introduction

Ray tracing methods have been said to be a useful way to predict radio propagation. Though the methods can be applied to any arbitrary room layout, lots of computation time is required to trace many propagation paths. The problem tends to prevent many engineers from using the ray tracing methods.

The ray tracing methods are classified as the imaging method or the launching method. Former method traces the ray paths between the transmitter location and the receiver location by producing its equivalent transmitter image with respect to the reflection object [1]. Latter method calculates amplitude at evaluation points by spreading rays from the transmitter location and trace its wavefront of the each ray.

The model uses the launching method has a potential to calculate fast compared with the imaging method in the case that lots of reflections occur. Because the launching method doesn't calculate for the evaluation points where the rays never launch. In spite of spreading many rays, not all rays launch the evaluation points where the rays should be launched. Though the method has been used as two-dimensional simulation, Seidel and Rappaport proposed the three-dimensional simulation technique [2]. The method represent discrete wavefront as geodesic domes, and a triangular icosahedron face is subdivided to generate the desired ray resolution. In spite of use of the sophisticated method, not all rays launch at accurate location.

The imaging method may be unrealistic when lots of reflections occur. The method considers all potential paths to reflect. Lots of computation time is required to distinguish existing propagation paths from the many candidates associated with the number of reflection objects and maximum reflection order. However, the imaging method has a advantage that the path between the transmitting point and the evaluation point is determined uniquely. It means that the rays launch the accurate locations on condition that the specular reflection is considered.

In this paper, a method which have a feature not to calculate for evaluation points where rays never launch is proposed. The computation time can be reduced because the calculation for evaluation points that the rays don't launch never be executed. In section **2.**, an investigation result of the probability that paths exist is shown. In section **3.**, the basic concept of the algorithm is presented. In the section, an example of the first-order reflection is used for heuristic understanding of the proposed method. In section **4.**, extension of the method for general reflection order is described. The performance evaluation is shown in section **5.**, and the conclusion is shown in section **6.**

2.   Investigation of the probability that paths exist

In the imaging method, the number of the potential paths becomes exponentially large as the number of the reflection objects becomes large. A room which has three reflection objects is considered as an example. The transmission point and the evaluation point are written as S and R, respectively, and the reflection point of three reflection object 1, 2, and 3 are defined as P_1, P_2, and P_3, respectively. In the case, the potential paths within second-order reflection can be described as : $S \rightarrow R, S \rightarrow P_1 \rightarrow R, S \rightarrow P_2 \rightarrow R, S \rightarrow P_3 \rightarrow R, S \rightarrow P_1 \rightarrow P_2 \rightarrow R, S \rightarrow P_1 \rightarrow P_3 \rightarrow R, S \rightarrow P_2 \rightarrow P_1 \rightarrow R, S \rightarrow P_2 \rightarrow P_3 \rightarrow R, S \rightarrow P_3 \rightarrow P_1 \rightarrow R, S \rightarrow P_3 \rightarrow P_2 \rightarrow R$. The number of the potential paths is 10. In general, the number of the potential paths p can be solved easily as (1)

$$p = 1 + w \sum_{i=0}^{n-1} (w-1)^i \qquad (1)$$

where w is the number of reflection objects, and n is the maximum reflection order. For example, a room which has 54 reflection objects is considered up to second-order reflection, 2,917 paths are determined for each evaluation point. If there are 10,000 evaluation points, 29,170,000 paths are determined for all evaluation points.

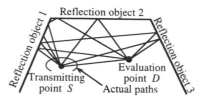

Figure 1   An example of ray paths

Not all potential paths have actual paths.   Therefore, the potential paths must be determined whether the paths exist or not.   Figure 1 is used for this example.   In the figure, there are three reflection objects, and there are 10 potential paths within second-order reflection.   However, there are only 7 actual paths in this case.

In order to investigate the probability that actual paths exist, ray tracing simulator which based on the imaging method is used.   An actual room layout which contains 54 reflection objects and 10,000 evaluation points is used as an example.   In this case, 237,053 actual paths are distinguished from 28,170,000 potential paths.   The probability that rays exist is less than one percent.   Several results show that less than few percents of the potential paths can be survived within second-order reflection.

3.   Basic concept of the proposed algorithm

According to the result in 2., the computation time can be reduced not to calculate for evaluation points where rays never launch.   To implement such feature, the calculation procedure is changed.   The conventional imaging method calculate by each evaluation point.   In contrast, the proposed method calculate by each object.

The reason of this change is described using Figure 2.   In this example, first-order reflection is used.   There is a evaluation area with a reflection object as depicted in Figure 2 (a).   And the evaluation area is filled with evaluation points which are equally spaced each other and they have same height [Figure 2 (b)].   A ray reflected by the object may launch at a point in one of the evaluation points as shown in Figure 2 (c).   Also rays are launched at evaluation points within a "illuminated area" on the evaluation plane shown in Figure 2 (d).   The illuminated area can be derived from the coordinates of the reflection object and the transmitter location.   However, rays reflected by the object never launch at evaluation points outside the illuminated area such as shown in Figure 2 (e).   Accordingly, it is enough to trace the rays reflected by the object at evaluation points within the illuminated area such as depicted in Figure 2 (f).   The concept can be applied to three-dimensional simulation.

114

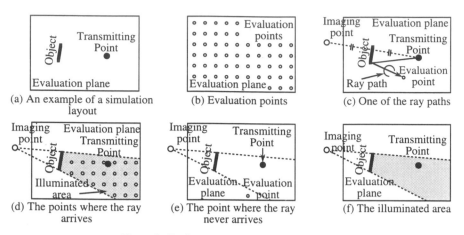

(a) An example of a simulation
layout

(b) Evaluation points

(c) One of the ray paths

(d) The points where the ray
arrives

(e) The point where the ray
never arrives

(f) The illuminated area

Figure 2   Basic concept of the proposed method

4. Extension of the method for general reflection order

In this section, extension of the method for general reflection order is discussed.   As an example, layout which contains two reflection objects shown in Figure 3 (a) is considered.   The figure also shows the illuminated area due to the reflection of the object 1.   Now a ray path through transmitting point, object 1, object 2, and evaluation point is considered.   A reflected rays due to the reflection object 1 illuminates a part of the object 2 as shown in Figure 3 (b).   If the same analogy of the concept ,previously discussed in section **3.** is applied the reflection due to the object 2, the illuminated area due to the object 1 reradiates the rays.   If three-dimensional arbitrary configuration is allowed, it may be possible that the illuminated area due to the object 1 is not covered with the object 2, such as depicted in Figure 3 (c).   In this case, the shape of the illuminated area on the evaluation plane may be a pentagon illustrated in Figure 3 (d).   In general, multiple reflections make its shapes more complex.   In order to avoid such cumbersome coordinate calculations and to be able to treat illumination area regardless of the reflection order, only last imaging point is used to solve the "approximated illumination area" on the evaluation plane such as shown in Figure 3 (e).   Not all evaluation points within the approximated illumination area have the actual ray path due to this modification.   Therefore, determination whether each point in the area has a actual path or not is needed.   The approximated illumination area can be solved by the coordinates of the last imaging point and the vertices of the last reflected object.   Three-dimensional approach is shown in Figure 4.

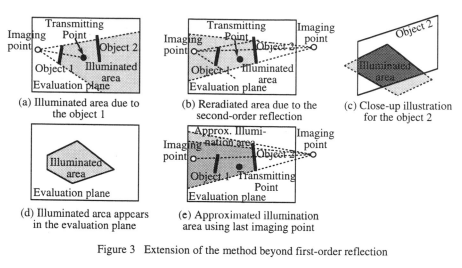

(a) Illuminated area due to the object 1

(b) Reradiated area due to the second-order reflection

(c) Close-up illustration for the object 2

(d) Illuminated area appears in the evaluation plane

(e) Approximated illumination area using last imaging point

Figure 3   Extension of the method beyond first-order reflection

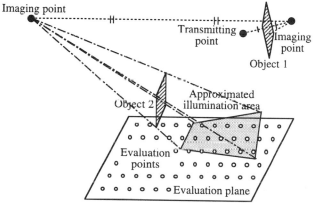

Figure 4   Three-dimensional implementation of the algorithm

5. Results

The radio propagation simulator incorporated above method is prepared to evaluate the performance. Lots of factors affect the computation time, for example, what computer is used, or how to make the program.   In this performance evaluation, Sun Microsystem's workstation SPARC station 10 is used in this experiment and C++ compiler is used for vector manipulations and for electrical magnetic calculations.

Line-of-sight, reflection, penetration are treated in this model. Fresnel's reflection coefficient is used to represent the loss of reflections and penetrations. Loss of reflection and penetration is decomposed into parallel component and perpendicular component.

116

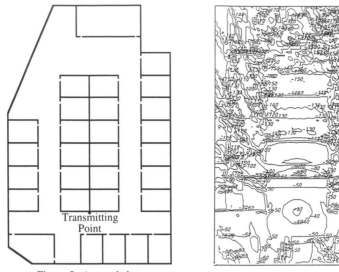

Figure 5   A sample layout

Figure 6   Calculation result of Figure 5

TABLE I   Computation time estimation of the proposed method

Condition of the evaluation			Computation time	
Layout	# of reflection objects	Max # of reflection	Proposed	Conventional
1	73	2	14 min	58 min
2	54	2	11 min	69 min

A sample layout used in this evaluation is shown in Figure 5, and its calculation result is shown in Figure 6.   In the figure, the values on the contour shows the loss with respect to the transmission point.   Computation time of some examples is tabulated in TABLE I.   In the table, the calculation is done over 10,000 evaluation points.   From the table, the computation time can be reduced to less than one-quarter of the time previously required.

6. Conclusion

The technique for the three-dimensional ray trace radio propagation simulation is proposed. The technique is based on the imaging method.   The coordinate relationship between the last reflected object and the last imaging point, which is derived from recursive image generation of the transmitting point, is used to approximate the illuminated area.   By incorporating these procedures into the conventional imaging method, the computation time can be reduced to less than one-quarter of the time previously required. Also, the accuracy of the proposed method is not degraded.

7. References

[1] John W. McKnown and R. Lee Hamilton, Jr : "A Ray Tracing as a Design Tool for Radio Networks", IEEE Network Magazine, pp. 27-30 (November 1991).
[2] Scott Y. Seidel and Theodore S. Rappaport : "A Ray Tracing Technique to Predict Path Loss and Delay Spread Inside Buildings", IEEE Globecom 92, pp. 649-653 (December 1992).

8. Appendix Calculation of illuminated area

This section describes how to solve the illuminated area using the coordinate relationships of the imaging point and the object. The scheme can be explained as following steps ;

(1) Calculate the parameters using the coordinates of the transmission (or imaging) point and vertices of the object by following equation (2).

$$
\begin{cases}
t_a = \dfrac{h - z_i}{z_u - z_i} \\[2mm]
t_b = \dfrac{h - z_i}{z_b - z_i} \\[2mm]
t_c = \dfrac{h - z_i}{z_c - z_i} \\[2mm]
t_d = \dfrac{h - z_i}{z_d - z_i}
\end{cases}
\tag{2}
$$

where z_a, z_b, z_c, z_d and z_i are the height of the vertices of the object and the imaging point, respectively, and h is the height of the evaluation plane.

(2) Calculate a state variable S. by using working variables (S_a, S_b, S_c, S_d) which are corresponded to the parameter (t_a, t_b, t_c, t_d), respectively. Regarding the parameter t_a,

    i)   if "$0 <= t_a <= 1$"   then S_a is set to "0"

    ii)   if "$1 < t_a$"   then Sa is set to "1",

    iii)   if "$t_a < 0$"   then Sa is set to "2."

The relationship between the configuration and the working variable S_a is illustrated in Figure 7.

118

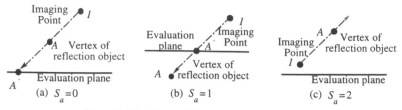

Figure 7 Working variable with respect to the vertex

Other working variables are set as same way. Then, the state variable S is defined as following equation (3),

$$S = 3^3 S_a + 3^2 S_b + 3^1 S_c + 3^0 S_d$$ (3)

(3) Rename the vertices of the reflection object, i.e., "A ← B", "B ← C", "C ← D", "D ← A". And repeat the procedures (1) thru (2) for four times.

(4) Find the minimum value S_{min}, which is the minimum value of the state values S. The combination of the minimum state variable is reduced to 6 cases, that is, it takes a value of 0, 4, 8, 40, 44, and 80. And calculate the illuminated area using the parameters (t_a, t_b, t_c, t_d) and the coordinates of the vertices (A, B, C, D) of the object.

There is a reason why the state variable which takes minimum value is used. For example in Figure 8, both (a) and (b) has same illuminated area, but the state variables of both cases are different. The value of the former case is "56", and the value of the latter case is "8." In order to reduce these ambiguity, the minimum state variable is used.

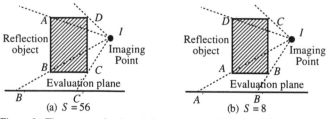

Figure 8 The reason why the minimum state variable should be found

The procedure to calculate the illuminated area is explained in details.

Figure 9 shows the case "$S_{min} = 0$." From the figure, the coordinates of the illuminated area is solved as equation (4).

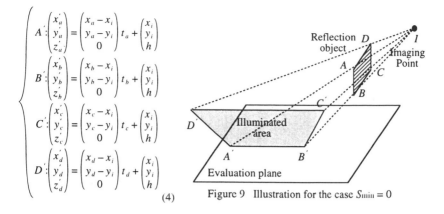

$$A: \begin{pmatrix} x'_a \\ y'_a \\ z'_a \end{pmatrix} = \begin{pmatrix} x_a - x_i \\ y_a - y_i \\ 0 \end{pmatrix} t_a + \begin{pmatrix} x_i \\ y_i \\ h \end{pmatrix}$$

$$B: \begin{pmatrix} x'_b \\ y'_b \\ z'_b \end{pmatrix} = \begin{pmatrix} x_b - x_i \\ y_b - y_i \\ 0 \end{pmatrix} t_b + \begin{pmatrix} x_i \\ y_i \\ h \end{pmatrix}$$

$$C: \begin{pmatrix} x'_c \\ y'_c \\ z'_c \end{pmatrix} = \begin{pmatrix} x_c - x_i \\ y_c - y_i \\ 0 \end{pmatrix} t_c + \begin{pmatrix} x_i \\ y_i \\ h \end{pmatrix}$$

$$D: \begin{pmatrix} x'_d \\ y'_d \\ z'_d \end{pmatrix} = \begin{pmatrix} x_d - x_i \\ y_d - y_i \\ 0 \end{pmatrix} t_d + \begin{pmatrix} x_i \\ y_i \\ h \end{pmatrix}$$

(4)

Figure 9   Illustration for the case $S_{min} = 0$

Figure 10 shows the case "$S_{min} = 4$." From the figure, the coordinates of the illuminated area is obtained as equation (5).

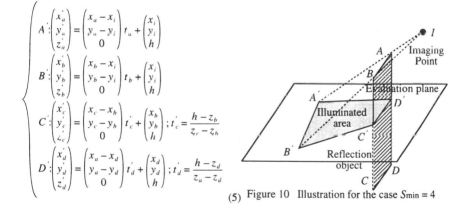

$$A: \begin{pmatrix} x'_a \\ y'_a \\ z'_a \end{pmatrix} = \begin{pmatrix} x_a - x_i \\ y_a - y_i \\ 0 \end{pmatrix} t_a + \begin{pmatrix} x_i \\ y_i \\ h \end{pmatrix}$$

$$B: \begin{pmatrix} x'_b \\ y'_b \\ z'_b \end{pmatrix} = \begin{pmatrix} x_b - x_i \\ y_b - y_i \\ 0 \end{pmatrix} t_b + \begin{pmatrix} x_i \\ y_i \\ h \end{pmatrix}$$

$$C: \begin{pmatrix} x'_c \\ y'_c \\ z'_c \end{pmatrix} = \begin{pmatrix} x_c - x_b \\ y_c - y_b \\ 0 \end{pmatrix} t'_c + \begin{pmatrix} x_b \\ y_b \\ h \end{pmatrix} ; t'_c = \frac{h - z_b}{z_c - z_b}$$

$$D: \begin{pmatrix} x'_d \\ y'_d \\ z'_d \end{pmatrix} = \begin{pmatrix} x_a - x_d \\ y_a - y_d \\ 0 \end{pmatrix} t'_d + \begin{pmatrix} x_d \\ y_d \\ h \end{pmatrix} ; t'_d = \frac{h - z_d}{z_a - z_d}$$

(5)

Figure 10   Illustration for the case $S_{min} = 4$

Figure 11 shows the case "$S_{min} = 8$." From the figure, the coordinates of the illuminated area is obtained as equation (6).

120

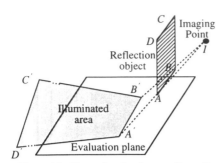

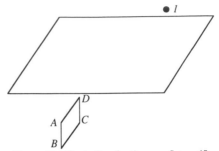

Figure 11   Illustration for the case $S_{min} = 8$

Figure 12   Illustration for the case $S_{min} = 40$

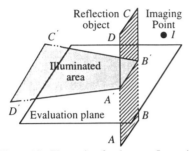

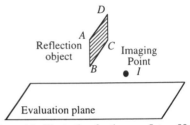

Figure 13   Illustration for the case $S_{min} = 44$

Figure 14   Illustration for the case $S_{min} = 80$

$$
\left\{
\begin{array}{l}
A: \begin{pmatrix} x'_a \\ y'_a \\ z'_a \end{pmatrix} = \begin{pmatrix} x_a - x_i \\ y_a - y_i \\ 0 \end{pmatrix} t_a + \begin{pmatrix} x_i \\ y_i \\ h \end{pmatrix} \\[20pt]
B: \begin{pmatrix} x'_b \\ y'_b \\ z'_b \end{pmatrix} = \begin{pmatrix} x_b - x_i \\ y_b - y_i \\ 0 \end{pmatrix} t_b + \begin{pmatrix} x_i \\ y_i \\ h \end{pmatrix} \\[20pt]
C: \begin{pmatrix} x'_c \\ y'_c \\ z'_c \end{pmatrix} = \begin{pmatrix} (x_c - x_b) t_e + x_b - x_i \\ (y_c - y_b) t_e + y_b - y_i \\ 0 \end{pmatrix} t_\infty + \begin{pmatrix} x'_b \\ y'_b \\ h \end{pmatrix} ; t_e = \dfrac{z_i - z_b}{z_c - z_b} \\[20pt]
D: \begin{pmatrix} x'_d \\ y'_d \\ z'_d \end{pmatrix} = \begin{pmatrix} (x_a - x_d) t_f + x_d - x_i \\ (y_a - y_d) t_f + y_d - y_i \\ 0 \end{pmatrix} t_\infty + \begin{pmatrix} x'_a \\ y'_a \\ h \end{pmatrix} ; t_f = \dfrac{z_i - z_d}{z_a - z_d}
\end{array}
\right.
$$

(6)

where t_∞ denotes the sufficient large value, for example, 10^5.

Figure 12 shows the case "$S_{min} = 40$." From the figure, the coordinates of the illuminated area is none.

Figure 13 shows the case "$S_{min} = 44$." From the figure, the coordinates of the illuminated area is obtained as equation (7).

$$\left\{ \begin{aligned} A': \begin{pmatrix} x'_a \\ y'_a \\ z'_a \end{pmatrix} &= \begin{pmatrix} x_a - x_d \\ y_a - y_d \\ 0 \end{pmatrix} t'_a + \begin{pmatrix} x_d \\ y_d \\ h \end{pmatrix} \; ; t'_a = \frac{h - z_d}{z_a - z_d} \\ B': \begin{pmatrix} x'_b \\ y'_b \\ z'_b \end{pmatrix} &= \begin{pmatrix} x_c - x_b \\ y_c - y_b \\ 0 \end{pmatrix} t'_b + \begin{pmatrix} x_b \\ y_b \\ h \end{pmatrix} \; ; t'_b = \frac{h - z_b}{z_c - z_b} \\ C': \begin{pmatrix} x'_c \\ y'_c \\ z'_c \end{pmatrix} &= \begin{pmatrix} (x_c - x_b) t_e + x_b - x_i \\ (y_c - y_b) t_e + y_b - y_i \\ 0 \end{pmatrix} t_\infty + \begin{pmatrix} x'_b \\ y'_b \\ h \end{pmatrix} \; ; t_e = \frac{z_i - z_b}{z_c - z_b} \\ D': \begin{pmatrix} x'_d \\ y'_d \\ z'_d \end{pmatrix} &= \begin{pmatrix} (x_a - x_d) t_f + x_d - x_i \\ (y_a - y_d) t_f + y_d - y_i \\ 0 \end{pmatrix} t_\infty + \begin{pmatrix} x'_a \\ y'_a \\ h \end{pmatrix} \; ; t_f = \frac{z_i - z_d}{z_a - z_d} \end{aligned} \right.$$

$$(7)$$

where t_∞ denotes the sufficient large value, for example, 10^5, same as the case "$S_{min} = 8$."

Figure 14 shows the case "$S_{min} = 80$." From the figure, the coordinates of the illuminated area is none.

Space vs. Polarization Diversity Gain in 2 GHz PCS 1900

by Paul Donaldson, Robert Ferguson, Eric Kmiec, and Robert Voss
MCI Telecommunications Corporation
901 International Parkway, Richardson, Texas 75081

1.0 Abstract

Measurements were taken during PCS 1900 field trials to quantify the difference in two branch diversity improvement between conventional space diversity of >10 wavelengths spacing and horizontal/vertical polarization diversity at   cell sites in the MCI test systems in Richardson, Texas and Washington, DC.   Data was taken with both the mobile unit antenna vertical, and with the mobile antenna tilted 45 degrees to the vertical to simulate the typical operating position of a handheld subscriber terminal. A simple evaluation method, using the inherent monitoring facilities contained within the GSM signal format, was developed and used during testing.   Under conditions representative of handheld operation, results indicate that base station receiver polarization diversity provided an attractive alternative to space diversity .

2.0 Introduction

MCI conducted various experiments over an 18 month period under a license granted by the Federal Communications Commission.[1]   These experiments encompassed a wide range of activities, including tests to determine appropriate maximum[2] and minimum size antenna configurations for various market deployment scenarios.   MCI investigated the use of "Smart" or intelligent antennas providing in excess of 23 dBi gain, physically large aperture size both horizontally and vertically (>30 wavelengths), and minimum size high performance configurations.

This paper describes the evaluation of the performance of two base station diversity receive configurations, one with conventional horizontal space diversity and   the other with horizontal/vertical (H/V) polarization diversity contained within the same physical antenna housing.   Base station frequencies were 1975 MHz transmit and 1895 MHz receive for all tests.

Tests were conducted in light industrial/suburban Richardson, TX and in urban Washington, DC environments.

Environmental and zoning considerations are a major factor in cell site location and construction. Polarization diversity reception requires only one half the number of antennas on the tower structure, and allows cell site antenna systems of significantly less visual impact. Additionally, no platform structure is required to provide spacing between antennas. Figure 1 shows a top down view of a three sector antenna configuration for both conventional space diversity and for a polarization diversity site.   Lee [3] suggests that at unobstructed base station sites desired diversity spacing between antennas should be >10 wavelengths for 0.7 correlation coefficient at an antenna height of 145 feet.

3.0 Discussion

Several previous experiments have reported results on polarization diversity under various scenarios. [4][5][6][7] Many experiments have focused on indoor propagation and determination of the best polarization (including circular) to provide reliable service.   For typical wall-mount base stations, obtaining the needed distance between horizontally spaced antennas becomes a logistics problem as both antennas cannot be conveniently mounted on the terminal itself.

In indoor tests, Ho and Rappaport [4] suggest that more signal depolarization will be seen at the receiver when an omni-directional mobile transmit antenna is used rather than a directional mobile antenna.   The antenna utilized on a PCS 1900 handset typically exhibits a 0 dBi omni-directional gain.

Smith and Neal [5] performed measurements at 900 MHz on an indoor CT2 system. They compared crossed dipoles (polarization diversity) vs. space diversity reception of mobile terminals using various antenna polarization configurations. They note that diversity gain was comparable with either receiving configuration and that the polarization of the mobile terminal had little effect on the key path loss or diversity parameters.

Vaughan [6] notes that the mechanism of decorrelation of the signals in each polarization is the multiple reflections occurring between the mobile and base antennas.   Some or all of the reflections cause coupling between the orthogonal polarizations. He suggests that after sufficient random reflections in a urban environment, the polarization state of the signal will be independent of the transmitted polarization. However, his results indicate that there is some

Figure 1 - Top Down View

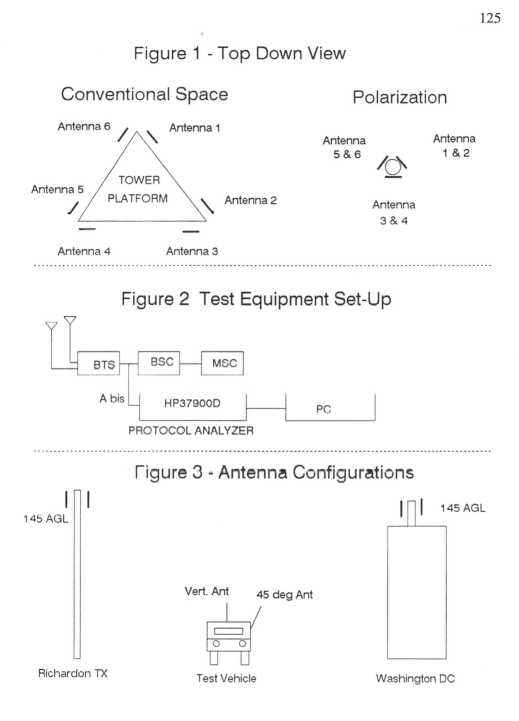

Conventional Space

Antenna 6   Antenna 1

TOWER
PLATFORM

Antenna 5   Antenna 2

Antenna 4   Antenna 3

Polarization

Antenna 5 & 6   Antenna 1 & 2

Antenna 3 & 4

Figure 2 Test Equipment Set-Up

BTS   BSC   MSC

A bis   HP37900D   PC

PROTOCOL ANALYZER

Figure 3 - Antenna Configurations

145 AGL

145 AGL

Vert. Ant   45 deg Ant

Richardon TX   Test Vehicle   Washington DC

dependence of the received polarization on the transmitted polarization present, even in urban environments.

McGladdery and Stapleton [7] discuss cross polarization coupling in an indoor environment and note that even when there is significant coupled energy, the amount of <u>correlated</u> cross coupling is expected to be small, providing good diversity performance.

Jakes [8] notes that polarization diversity is one of several methods useful for improving system performance. He notes that signals transmitted on two orthogonal polarizations in the mobile environment exhibit uncorrelated fading statistics.

In a multi-path environment, there will be reflections causing loss of polarization sense. Interestingly, Jakes notes that at the base station, the received signal strengths of simultaneous horizontal and vertical path loss are within 3 dB of each other 90% of the time; this behavior is independent of base station antenna height. This contrasts significantly with the result derived from vector addition of direct and ground reflected waves. Jakes suggests that the most important factor in determining the average signal strength is the total amount of power coming from all directions.

Eggers [9] performed tests at 900 MHz by varying both mobile and base antenna configurations between vertical and horizontal polarization. In urban environments, he reports XPD of 4 dB.

4.0 MCI Test Configuration

The test system consisted of Northern Telecom/Matra PCS 1900 equipment configured in a complete GSM network. It is important to note that the base station receivers use two-branch **maximal ratio combiners**. The system layout is shown in Figures 2 and 3. The antenna type used in the test was the Northern Telecom CellPlus (TM) which provides both vertical and horizontal diversity reception. The antenna housing contains two independent antennas, one with vertical and one with horizontal polarization. Antenna gain is 18.5 dBi with a pattern 3 dB beamwidth of 60 degrees wide by 6 degree high. For space diversity, the test used the vertically polarized antenna in two CellPlus antennas which were spaced seven feet horizontally apart.

The base station antennas were about 145 feet AGL in each city. In the urban test area of Washington, DC, this height is about 15 feet above the rooftops of most nearby buildings. Richardson antennas were located in a light industrial area upon a self-supporting tower with no nearby objects of comparable height. However, some buildings up to 20 stories high were present in the test area.

In each city, a single drive route was used. The Washington DC drive route was through the urban area with few locations line-of-sight to the base antennas. All points were within 1.2 miles of the base station. The Richardson drive route included highway, light industrial and residential segments within 2.6 miles of the base station. Significant portions of the route were line-of-sight or near line-of-sight.

Our test utilized two mobile antenna system configurations. The first system consisted of a vertically mounted antenna, which would then require the local environment to provide sufficient reflections (and phase variations) to see any significant energy received by the horizontally polarized receive antenna. In the second case, the transmit antenna was oriented at 45 degrees, at varying orientations toward the base site. (Conventional polarization diversity would assume that equal energy was transmitted both horizontally and vertically intentionally.) In our tests, we simulated the performance likely to be seen by a handset in the normal use position.

5.0 Test Methodology

Practical considerations cause 'real world' (non-laboratory) evaluations to be a compromise of the ideal comparison procedure. The technique developed for characterizing the diversity gain represents an approach that minimizes the need to continually calibrate system reference levels and for specialized test equipment. The downlink path is used as a common reference for each of the test runs. The downlink path does not change between runs and provides an 'intermediate reference' to which uplink test cases can be compared.

Data was collected for three base receive antenna configuration cases: (1) non-diversity, (2) space diversity, and (3) H/V polarization diversity. The gain achieved in each configuration was evaluated in a dynamic propagation environment with a reference performance BER threshold range chosen to represent actual service expectations. Desirable system enhancement is most significant in the operating region where performance of the radio link is marginal. Uplink improvement that allows less subscriber terminal power (and thus less interference) is useful at all signal levels.

A standard PCS 1900 handset mobile unit was installed in a vehicle. A vertically mounted quarter wave antenna on a self-contained ground plane was located above the vehicle roof. The second antenna was a vertical coaxial sleeve dipole mounted at a 45 degree angle above the vehicle. This second antenna orientation is suggested as more representative of handheld operation next to the head where the unit is often tilted at a 30-45 degree angle.

A GPS location system was used in the vehicle, and the position recorded during each run on a laptop PC. Accurate time correlation between position data and radio performance data allowed files to be combined for analysis.

The technique can be summarized as follows:
1. Set an initial uplink diversity configuration on the base station transceiver.
2. Drive a predetermined route, collecting uplink and downlink signal and performance data. Record mobile location data on PC in vehicle.
3. Specify an uplink performance range that will be used to make configuration comparisons.
4. At points in the drive run which showed error rates in the selected uplink performance range, calculate the mean downlink signal level.
5. Repeat for each uplink configuration (i.e. change antenna set up at base).
6. Process data and perform statistical analysis on link performance parameters.
7. Compare the mean downlink signal levels (coincident with the chosen uplink performance range) for each configuration, using dB differences directly to compare uplink configuration performance.

Advantages of this approach are:
1. Only one standard base station and one standard mobile unit are required for all tests.
2. The single set of equipment reduces calibration requirements and minimizes the likelihood of unidentified (perhaps subtle) differences from run to run. Downlink performance can be used as a test of consistency.
3. The reported uplink signal level (which may not be precisely defined in a radio system with combining) is not utilized as a determining factor in the computation of system gain

improvement. The mean downlink propagation loss calculated from measurements at many points is expected to closely estimate the mean uplink propagation loss at the same points. The uplink and downlink frequencies differed by 80 MHz.

6.0 Base Station Implementation

The PCS 1900 system measures signal level and error rate in both the uplink and downlink direction and reports it to the base station controller every 480 milliseconds [10]. The quality of the transmission paths is used in the handoff algorithm, for normal system maintenance and operation, and for system coverage verification.

In GSM, the Quality Band (see Table 1) represents the estimated error probability before channel decoding, averaged over a full set of 104 TDMA frames. The following table presents the RX Quality (RXQUAL_FULL) levels reported by the system.[11]

Table 1 - RXQUAL_FULL

Quality Band	Equivalent BER	Assumed Value
0	Less than 0.2%	0.14%
1	0.2 to 0.4%	0.28%
2	0.4 to 0.8%	0.57%
3	0.8 to 1.6%	1.13%
4	1.6 to 3.2%	2.26%
5	3.2 to 6.4%	4.53%
6	6.4 to 12.8%	9.05%
7	greater than 12.8%	18.10%

An external protocol analyzer connected to the base station provided the data decoding required to monitor the signal level and quality of the links in both directions. A simple data logging system storing this information to disk or tape is then used to complete the measurement setup. The recording system is shown in Figure 2. Tests were conducted without discontinuous transmission, mobile power control, or frequency hopping.

7.0 Results of Tests

Each run consisted of approximately 2000 data points. For each diversity configuration, a test run was completed around the same drive route. Representative downlink data is presented in Table 2. The downlink mean signal level for each downlink quality band is shown for a set of

Figure 4

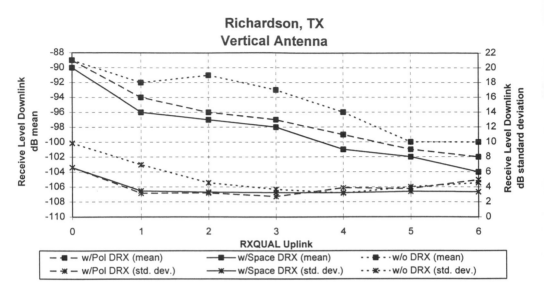

Figure 5

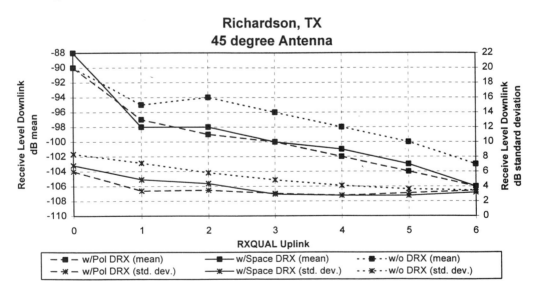

three runs with each mobile antenna orientation. Consistency of the downlink path signal level within each set is used to verify validity of the uplink results.

Table 2 - Sample Test Downlink Data - Richardson

Downlink Level dB Mean (dBm) vs. Downlink Quality Band								
Antenna Orient.	Quality Band							
	0	1	2	3	4	5	6	7
1/4 wave Vertical	-86.0	-92.0	-93.6	-95.8	-98.8	-101.2	-103.8	-108.3
	-87.0	-91.7	-93.0	-94.7	-98.1	-100.8	-103.8	-104.2
	-87.7	-93.5	-94.1	-96.0	-98.7	-101.1	-104.1	-106.6
1/2 wave Dipole @ 45°	-95.7	-96.5	-96.5	-97.8	-99.3	-101.3	-103.6	-106.8
	-89.2	-93.1	-94.4	-96.5	-98.9	-101.2	-103.8	-106.7
	-95.3	-96.0	-96.7	-98.5	-99.7	-101.6	-104.2	-107.6

Figures 4, 5, 6 and 7 show sample results from runs with the three diversity configurations for both vertical and 45 degree mobile antenna orientations. Figures 4 and 5 are for Richardson and Figures 6 and 7 are for the Washington tests. The figures plot the mean downlink level observed to achieve uplink Quality Bands 0-6 for a test run in each configuration. The standard deviation of the downlink signal is also shown on each figure.

The Quality Bands 2-4 were chosen to evaluate the diversity gain improvement. Bands 0-1 represent good performance, while Bands 5-7 do not provide adequate system performance to meet service objectives.

Table 3 presents the computed average dB diversity improvement seen for all data points with uplink in Quality Bands 2-4 for each configuration and mobile antenna orientation.   As discussed previously, mean downlink signal level differences have been used to estimate these gains.

132

Figure 6

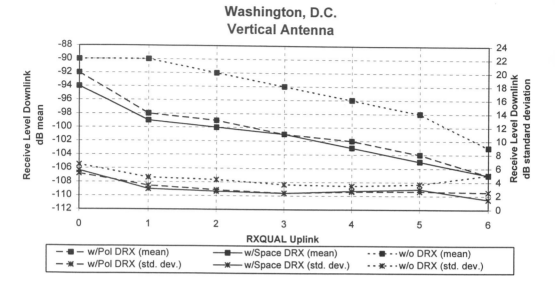

Figure 7

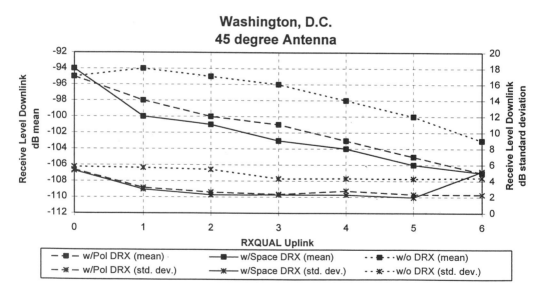

Table 3 - Test Results: Diversity Gain

Run Set	Mobile Ant. Orientation	Polarization Diversity	Space Diversity
Richardson - Set #1	Vertical	5.2 dB	5.0 dB
Set #2	Vertical	3.5 dB	5.2 dB
Set #3	45 degree	4.4 dB	3.6 dB
Washington - Set #1	Vertical	6.6 dB	7.3 dB
Set #2	Vertical	6.9 dB	7.3 dB
Set #3	45 degree	4.8 dB	6.5 dB
Set #4	45 degree	5.4 dB	6.7 dB

Although the same identical drive route was used for all tests within a city, traffic patterns varied from run to run.

8.0 Conclusions

A simple technique to compare diversity configuration gain has been described. By using the reported downlink signal as a reference, uplink performance in various configurations can be compared and a decibel difference found using simple equipment configurations on a standard system. For typical handset antenna orientations, polarization diversity at the base station receiver compared favorably with space diversity.

9.0 Acknowledgment

MCI wishes to thank NTI and MATRA personnel, Thierry Prevost, Joel Guyonnaud, Antoine Clatot, who assisted with MCI's field trials in Richardson, TX and Washington, DC with the PCS 1900 test system.

134

10.0 References

[1] Experimental Licenses KQ2XJO and KQ2XLE. MCI has filed quarterly reports to the FCC. MCI conducted both QCDMA and PCS 1900 tests. This analysis deals only with selected PCS 1900 test runs.

[2] Keller, C.M., and Brown, M.C., "Propagation Measurements for 1.9 GHz High-Gain Antenna Design", to be presented I.E.E.E. V.T.G. 1995. This work was prepared under a Cooperative Research and Development Agreement between MIT Lincoln Labs and MCI.

[3] Lee, W.C.Y., *Mobile Communications Engineering*, McGraw-Hill Book Co.

[4] Ho, C.M Peter, and Rappaport, Theodore S. , "Effects of Antenna Polarization and Beam Pattern on Multipath Delay Spread and Path Loss in Indoor Obstructed Wireless Channels"

[5] Smith, M.S., and Neal, L, "A Comparison of Polarisation and Space Diversity for Indoor Propagation at 900 MHz", ICUPC 1993

[6] Vaughan, Rodney, "Polarization Diversity in Mobile Communications", *IEEE Transactions of Vehicular Technology*, Vol 39, No 3, August 1990

[7] McGladdery, W. A., and Stapleton, S., "Investigation of Polarization Effects in Indoor Radio Propagation", IEEE ICWE 1992

[8] Jakes, William, C., *Microwave Mobile Communications*, IEEE Press

[9] Eggers, Patrick, Toftgard, Jorn, and Oprea, Alex, "Antenna Systems for Base Station Diversity in Urban Small and Micro Cells", *IEEE Journal on Selected Areas in Communications, Vol 11, No. 7, Sept. 1993*

[10] Mouly, Michel, and Pautet, Marie-Bernadatte, *The GSM System for Mobile Communications*

[11] ETSI GSM Recommendation, Section 5.08 (v4.6.0), para 8.2.4, July 1993

Minimization of Outage Probability in Cellular Communication Systems by Antenna Beam Tilting

Josef Fuhl and Andreas F. Molisch

Institut für Nachrichtentechnik und Hochfrequenztechnik
Technische Universität Wien; Vienna, Austria
Gusshausstrasse 25/389, A-1040 Wien, Austria
Tel.: (+43 1) 58801/3546 Fax.: (+43 1) 587 05 83

Abstract

This paper analyses the effect of antenna beam tilting on the outage probability of a cellular communication system. Different cell geometries, antenna patterns, carrier-to-interference ratios, and carrier-to-noise ratios are adressed.

Our results show that determination of the optimum tilt angle is important especially for interference limited cells. For an antenna height in excess of 50m, missing the optimum tilt angle by 1° leads to an increase in the outage probability by a factor of up to seven.

1 Introduction

In wireless communication systems the same frequency channels have to be used many times due to limited spectrum resources. This reuse causes cochannel interference, which is one of the major sources of performance degradation. In order to obtain larger carrier-to-interference ratios (CIR), and to minimize delay spread in TDMA systems, the vertical antenna pattern can be tilted downwards by some degrees [1], [2]. This results in a concentration of emitted power to the desired cell and in a decrease of power emitted to adjacent cells. On the other hand, too large a tilt can decrease the carrier-to-noise ratio (CNR) at the cell fringe. Consequently, there is an optimum tilt angle that minimizes the outage probability [3]. In this paper, we determine this optimum tilt angle as function of CNR, CIR, antenna pattern, and antenna height.

Most of the treatments published earlier consider only the CIR as the quantity relevant for determination of the outage probability and the optimum tilt angle [4]. We consider the outage probability due to both noise and interference in a basic cell configuration first with one interfering base station, and show how this can be extended to N interfering base stations. For the outage probability we derive a formula containing only one numerical integral. From this expression we determine the optimum tilt angles for different antenna dimensions (number of elements), antenna heights, CNR, CIR, and reuse factors including the special cases of purely noise limited and purely interference limited systems. Furthermore, we give analytical approximations for the optimum tilt angle and compare them to the numerical results.

Our results show that the optimum tilt angle is near the *noise limit angle*[1] for small antenna heights and approaches the *interference limit angle*[2] for large antenna heights. The functional relationship between the optimum tilt angle and the antenna height is given by a square–root–like function, with CNR, CIR, number of antenna elements, and the reuse factor as parameters.

Our paper is organized as follows. Section 2 reviews the propagation model used. In Section 3 we give an analytical expression for the outage probability for a cellular communication system. Section 4 shows the optimum tilt angle for different cell geometries and its dependence on the system parameters. In Section 5 a polynomial fit for the optimum tilt angle is given. Conclusions are presented in Section 6.

2 Propagation Model

In mobile radio the area mean power $m_d(r)$ of the signal received from a mobile subscriber at distance r from the base station can be modelled as [5]

$$m_d(r) = 10 \log(P_t r^{-\alpha}) + G(\theta_{user}(r)), \tag{1}$$

where P_t is the transmitted power, α is in the range between 2 and 5, and $G(\theta_{user}(r))$ is the gain of the base station antenna in the direction of the user. The direction of the user, $\theta_{user}(r)$, is given by

$$\theta_{user}(r) = \arctan(\frac{H_B}{r}), \tag{2}$$

where H_B is the height difference between the base station antenna and the mobile station antenna.

Assuming shadowing (slow fading) superimposed on (1), the local mean power X_s is log–normally distributed about the area mean power

$$p(X_s) = \frac{1}{\sqrt{2\pi}\sigma} e^{-\frac{(X_s - m_d(r))^2}{2\sigma^2}}, \tag{3}$$

where σ is the logarithmic standard deviation of the shadowing.

This propagation model is valid for both signal and interference.

If a mobile receives signals from more than one base station, the shadowing of these signals may be correlated with a correlation coefficient ρ [6]. As a measure for this correlation the angular difference between the directions of arrival can be used. We define the angle difference for two impinging signals from different base stations as the angle difference between two lines connecting the mobile station with the corresponding base stations (mod 180°). Decreasing correlation with increasing angle difference can be observed [6].

---

[1] which is the optimum tilt angle derived by considering noise only
[2] which is the optimum tilt angle derived by considering cochannel interference only

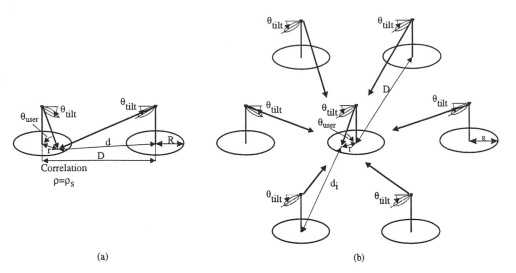

Figure 1: Cell structures. (a) Simple frequency reuse model, (b) Hexagonal cell frequency reuse model

3   Calculation of Outage Probability

In the following derivation we consider only the influence of shadowing, and do not take into account the effect of Rayleigh fading.

Figure 1 shows the basic cellular structures considered in this paper. In Figure 1a) the simplest model with one interfering base station is shown, Figure 1b) shows a commonly used structure with 6 interfering cells.

Let X_s and X_u be the local means of the desired and the undesired signal, respectively. If there is only one interfering base station, then the undesired signal is characterized by its area mean $m_u(d)$ and its standard deviation σ. If there are N (uncorrelated) interfering signals, all with the same area mean $m_{u,1}(d)$ and the same variance σ, then the resulting total interfering signal has a local mean [7], [3]

$$m_u(d) = m_{u,1}(d) + 10 \log(N), \tag{4}$$

and a standard deviation

$$\sigma_u = \left[\log_e \left(1 + \frac{e^{\left(\frac{\sigma}{10 \log_{10} e}\right)^2} - 1}{N} \right) \right]^{1/2} \times 10 \log_{10} e. \tag{5}$$

The assumption of equal area means of the N interfering signals implies that the distances of the mobile to each base station and the propagation conditions are in the same order.

X_s and X_u are jointly normal and their probability density function (pdf) is given by [9]

$$p(X_s, X_u, \rho_u) = \frac{1}{2\pi\sigma\sigma_u\sqrt{1-\rho_u^2}} e^{-\frac{1}{2(1-\rho_u^2)}\left[\frac{(X_s-m_s(r))^2}{\sigma^2} - \frac{2\rho_u(X_s-m_s(r))(X_u-m_u(d))}{\sigma\sigma_u} + \frac{(X_u-m_u(d))^2}{\sigma_u^2}\right]}, \quad (6)$$

with $\rho_u = \rho$ for $N = 1$ and $\rho_u = 0$ for $N > 1$ [6]. Let γ and λ denote the carrier–to–noise ratio (CNR) and the carrier–to–interference ratio (CIR) of the local mean, respectively. Then $\gamma = X_s$ and $\lambda = X_s - X_u$. With this transformation (6) becomes

$$p(\gamma, \lambda, \rho_u) = \frac{1}{2\pi\sigma\sigma_u\sqrt{1-\rho_u^2}} e^{-\frac{1}{2(1-\rho_u^2)}\left[\frac{(\gamma-m_s(r))^2}{\sigma^2} - \frac{2\rho_u(\gamma-m_s(r))(\gamma-\lambda-m_u(d))}{\sigma\sigma_u} + \frac{(\gamma-\lambda-m_u(d))^2}{\sigma_u^2}\right]}. \quad (7)$$

The *local outage probability* is then defined by

$$P_{out} = P[\gamma < \gamma_{th} \quad \text{or} \quad \lambda < \lambda_{th}], \quad (8)$$

where γ_{th} and λ_{th} are the threshold values for CNR and CIR of the local mean, respectively. In accordance to [8], [3], the local outage probability can be split into

$$
\begin{aligned}
P_{out} &= P_{out}^N + P_{out}^I - P_{out}^{N\wedge I} \\
&= P[\gamma < \gamma_{th}] + P[\lambda < \lambda_{th}] - P[\gamma < \gamma_{th} \quad \text{and} \quad \lambda < \lambda_{th}] \\
&= \int_{-\infty}^{\gamma_{th}} \int_{-\infty}^{\infty} p(\gamma, \lambda; \rho_u)\, d\lambda\, d\gamma + \int_{-\infty}^{\infty} \int_{-\infty}^{\lambda_{th}} p(\gamma, \lambda; \rho_u)\, d\lambda\, d\gamma - \int_{-\infty}^{\gamma_{th}} \int_{-\infty}^{\lambda_{th}} p(\gamma, \lambda; \rho_u)\, d\lambda\, d\gamma, \quad (9)
\end{aligned}
$$

where P_{out}^N is the local outage probability due to noise only, P_{out}^I is the outage probability due to interference only, and $P_{out}^{N\wedge I}$ is the local outage probability due to noise and interference. Equation (9) can also be written as

$$
\begin{aligned}
P_{out} &= 1 - P[\gamma > \gamma_{th} \quad \text{and} \quad \lambda > \lambda_{th}] \\
&= 1 - \int_{\gamma_{th}}^{\infty} \int_{\lambda_{th}}^{\infty} p(\gamma, \lambda; \rho_u)\, d\lambda\, d\gamma, \quad (10)
\end{aligned}
$$

where

$$P[\gamma > \gamma_{th} \quad \text{and} \quad \lambda > \lambda_{th}] = \frac{1}{2\sqrt{2\pi}\sigma}\left[\sqrt{\frac{\pi}{2}}\sigma\text{erfc}\left(\frac{\gamma_{th}-m_s(r)}{\sqrt{2}\sigma}\right) - \int_{\gamma_{th}}^{\infty} e^{-\frac{(\gamma-m_s(r))^2}{2\sigma^2}} \text{erf}\left(\frac{\lambda_{th}}{\sigma_u\sqrt{2(1-\rho_u^2)}} + \gamma\frac{\rho_u\sigma_u-\sigma}{\sigma\sigma_u\sqrt{2(1-\rho_u^2)}} + \frac{m_u(d)\sigma-\rho_u m_s(r)\sigma_u}{\sigma\sigma_u\sqrt{2(1-\rho_u^2)}}\right) d\gamma\right], \quad (11)$$

and erfc(.) (erf(.)) are given by

$$\text{erfc}(x) = 1 - \text{erf}(x) = \frac{2}{\sqrt{\pi}} \int_x^{\infty} e^{-t^2}\, dt. \quad (12)$$

4 Optimum Tilt Angle

The optimum tilt angle of the vertical antenna pattern is defined as the tilt angle that results in the smallest outage probability

$$\theta_{tilt,opt} = \min_{\theta_{tilt}} \{P_{out}\}. \quad (13)$$

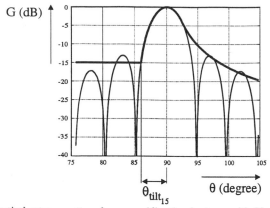

Figure 2: Vertical antenna pattern for an omnidirectional antenna with $M = 12$ elements with element spacing $d = \lambda$. The heavy line represents the approximation of the antenna pattern used for the calculations.

For our simulations we used a cell radius of $R=1.5$ km, a distance of the mobile to its serving base station of $r = R = 1.5$ km, a distance between the base stations (cells) of $D=6$ km, $N = 1$, a distance of the mobile to the interfering base station of $d = D - R = 4.5$ km, $\gamma_{th} = \lambda_{th} = 7$ dB. If not mentioned otherwise we assumed $\rho_u = 0$ and $\sigma = 6.5$dB. As antenna we used an omnidirectional antenna with $M = 12$ antenna elements with spacing $d = \lambda$. The antenna pattern is shown in Figure 2. We approximated the antenna pattern in the interference region by a -15 dB–line and in the coverage region by its envelope, since due to multipath propagation the nulls are filled up to a certain extent [10], [11].

We calculate the outage probability only for the cell fringe, since this is usually the most critical region of the cell.

4.1   Outage Probability versus Antenna Height

Figure 3 shows the outage probability versus the tilt angle with the height of the base station antenna and the reuse factor $U = D/R$ as parameters. Γ and Λ are given by

$$
\begin{aligned}
\Gamma &= 10\log_{10}(P_t r^{-\alpha}) + G_0 - \gamma_{th} \\
\Lambda &= 10\log_{10}(P_t r^{-\alpha}) - 10\log_{10}(P_t d^{-\alpha}) - 10\log_{10}(N) - \lambda_{th}.
\end{aligned}
\tag{14}
$$

The outage probability decreases with increasing height of the antenna. The influence of the reuse factor is not as large as that of the antenna height. With increasing reuse factor the outage probability decreases, even if the same power levels are present at the mobile.

Figure 4 shows the optimum tilt angle as function of the antenna height with Γ and Λ as parameters. From this figure two boundary curves can be identified, the *noise limit angle* $\theta_{tilt,opt}^N$ given by

$$
\theta_{tilt,opt}^N = \min_{\theta_{tilt}}\{P_{out}^N\} = \arctan(\frac{H_B}{r}) \approx \frac{H_B}{r}
\tag{15}
$$

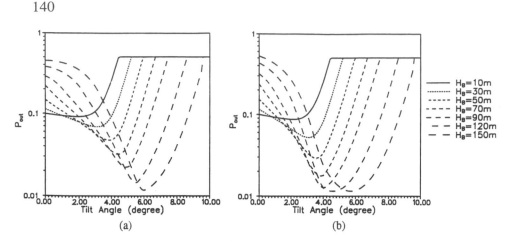

Figure 3: Outage probability P_{out} versus the tilt angle θ_{tilt} for (a) $D/R = 4$ and (b) $D/R = \infty$. $\Gamma = 22$dB, $\Lambda = 19$dB

and the *interference limit angle* $\theta^I_{tilt,opt}$ given by

$$\theta^I_{tilt,opt} = \min_{\theta_{tilt}}\{P^I_{out}\} = \begin{cases} \theta_{tilt_{15}} + \arctan(H_B/d) \approx \theta_{tilt_{15}} + H_B/d & : & H_b \leq H_{B,th} \\ \arctan(H_B/r) \approx H_B/r & : & H_b > H_{B,th} \end{cases}, \quad (16)$$

where $\theta_{tilt_{15}}$ is the angle difference (in radian) between the angle where the radiation pattern in the interference region reaches the -15 dB level and the angle of the main beam.

If the height of the base station antenna is above the threshold $H_{B,th}$ given by

$$H_{B,th} = r\frac{U-1}{U-2}\theta_{tilt_{15}}, \quad (17)$$

then $\theta^N_{tilt,opt}$ and $\theta^I_{tilt,opt}$ are the same. Figure 5 shows the influence of the correlation coefficient ρ_u. As the correlation increases, the outage probability decreases and optimum tilt angle approaches the noise limit angle $\theta^N_{tilt,opt}$. This is due to the fact that if the shadowing of the wanted and the unwanted signals is strongly correlated, the corresponding envelopes fluctuate in the same manner. The influence of P^I_{out} on the optimum tilt angle $\theta_{tilt,opt}$ is becoming less and less due to the fact that the CIR is almost constant for every specific shadowing situation.

4.2 Sensitivity Analysis

A sensitivity analysis of the optimum tilt angle is done in Figures 6 and 7. Figure 6 shows the influence of a tilt angle error of 1° on the outage probability P_{out}. The most critical situation is given for the interference limited case and large antenna heights, where an error of 1° can raise the outage probability by a factor of up to seven.

In Figure 7 the influence of different antennas ($M = 8$, $M = 12$, and $M = 16$ elements) is shown. If the system is not purely noise limited, a larger antenna decreases the outage

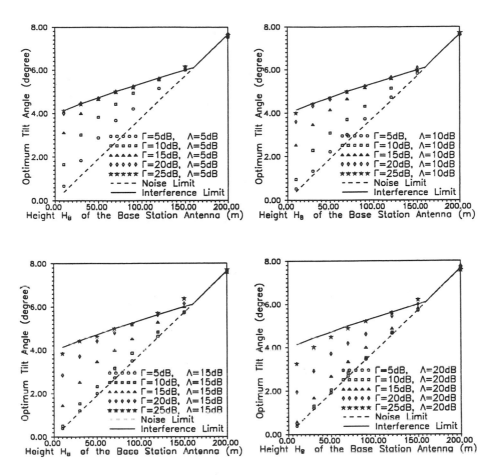

Figure 4: Optimum tilt angle $\theta_{tilt,opt}$ versus the antenna height $(D/R = 4)$

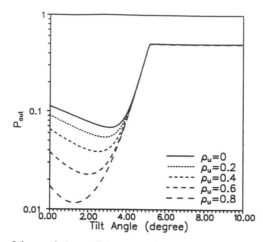

Figure 5: Influence of the correlation coefficient ρ_u on the optimum tilt angle $\theta_{tilt,opt}$ ($D/R = 4$, $\Gamma = 15$dB, $\Lambda = 12$dB, $H_B = 30m$).

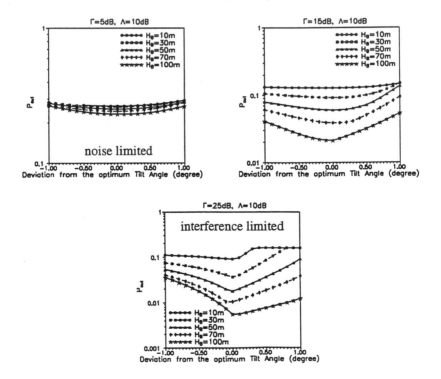

Figure 6: Influence of a tilt angle error on the outage probability for different CNR and CIR ($D/R = 4$)

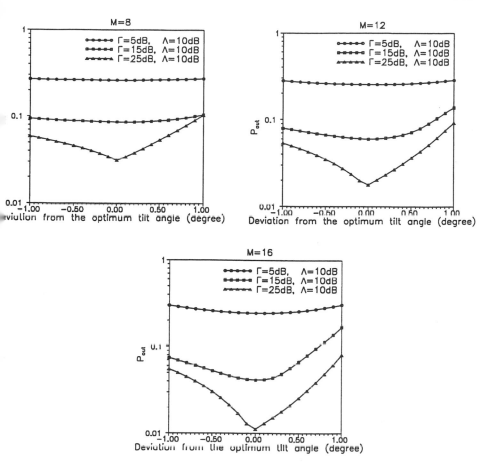

Figure 7: Influence of different antenna sizes and a tilt angle error on the outage probability ($D/R = 4$, $H_B = 50m$)

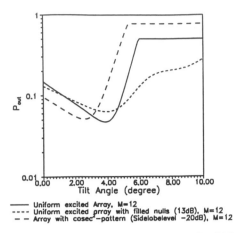

Figure 8: Influence of different antenna patterns on the outage probability ($D/R = 4$, $\Gamma = 15$dB, $\Lambda = 12$dB, $H_B = 50m$).

probability due to its sharper rolloff of the main beam. This decreases the radiation emitted to adjacent cells compared to the amount delivered to the own cell. Also, the sensitivity of the outage probability with respect to tilt ange errors increases with increasing CNR (at constant CIR) and increasing antenna size.

Figure 8 shows the influence of different antenna patterns on the outage probability. It shows that a cosec²–shaped pattern with −20dB sidelobes increases the outage probability because its main beam is broader than that of the corresponding linear array. However, a cosec²–shaped pattern is beneficial for users within the cell, and would lead to a decrease of the outage probability if we average over the whole cell. Filled nulls increase the outage probability at the cell fringe because of the increased sidelobe level.

5 Approximations for the Optimum Tilt Angle

In order to be able to give a closed–form equation for the optimum tilt angle $\theta_{tilt,opt}$ we fit it as

$$\theta^f_{tilt,opt} = \theta^f_{tilt,opt}(\Gamma, \Lambda, H_B, M, \ldots) = \theta^f_{tilt,opt}(\Gamma) + \theta^f_{tilt,opt}(\Lambda) + \theta^f_{tilt,opt}(H_B) + \theta^f_{tilt,opt}(\Gamma - \Lambda), \quad (18)$$

with

$$\theta^f_{tilt,opt}(\Gamma) = \sum_{i=1}^{N_a} a_i \left(\frac{H_B}{r} \Gamma \right)^i$$

$$\theta^f_{tilt,opt}(\Lambda) = \left(\frac{H_B}{d} + \theta_{tilt_{15}} \right) \sum_{i=1}^{N_a} b_i \left[\left(\frac{H_B}{d} + \theta_{tilt_{15}} \right) \Gamma \right]^i$$

$$\theta^f_{tilt,opt}(H_B) = \sum_{i=1}^{N_a} c_i H_B^i$$

i	a_i	b_i	c_i	e_i
1	1.1823	1.3555	0.5908	-3.118
2	-1.1276	-0.5886	-0.0363	0.1886
3	0.2913	0.0869	0.0012	1.4454

Table 1: Coefficients for the polynomial fit for $N_a = 3$

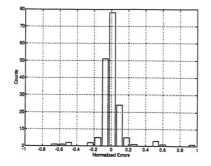

Figure 9: Histogram of the normalized errors ϵ for a fit with all data.

$$\theta^f_{tilt,opt}(\Gamma - \Lambda) = \sum_{i=1}^{N_a} e_i \left\{ \left[\frac{H_R}{r} - \left(\frac{H_R}{d} + \theta_{tilt_{15}} \right) \right] (\Gamma - \Lambda) \right\}^i. \qquad (19)$$

The fitting coefficients $a_i, b_i, c_i,$ and e_i were derived from a totally–least–squares fit. Table 1 shows the values of the coefficients for $N_a = 3$. Figure 9 shows a histogram of the normalized errors defined by

$$\epsilon = \frac{\theta_{tilt,opt} - \theta^f_{tilt,opt}}{\theta_{tilt,opt}}. \qquad (20)$$

The standard deviation for the normalized errors is given by $\sigma_{res} = 15.79\%$. If we exclude the optimum tilt angles for $H_B = 10$m from our sample, we obtain a standard deviation of $\sigma_{res} = 6.64\%$. (The exclusion of these values is justified by the observation that the absolute value of the outage probability for $H_B = 10$m is nearly independent of the tilt angle when close to optimum (see Figures 3, 6, and 7). Therefore our fitting coefficients are in first order also valid for $H_B = 10$m.)

6   Conclusions

We have analyzed the influence of the tilt angle of the antenna beam on the outage probability on the cell fringe of a cellular communication system.

Our analysis shows that, in noise limited systems, the outage probability is not sensitive to the tilt angle. However, in interference limited systems determination and realization of the optimum tilt angle is a must to achieve best system performance with given equipment.

7 Acknowledgment

Support of this work by the Austrian PTT is gratefully acknowledged. The views expressed in this paper are those of the authors and do not necessarily reflect the views within the Austrian PTT.

We want to thank Prof. Ernst Bonek for stimulating discussions.

References

[1] Y. Yamada, and M. Kijima. "Low Sidelobe and Tilted Beam Base Station Antennas for Small-Cell Systems", IEEE Trans. AP–S, pp. 138–141, 1989.

[2] J.-F. Wagen, J.-P. de Weck, J.-F. Zürcher, and J. Sanford. "Time Dispersion Measurements Using a SSFIP Base Station Antenna", Proc. 42th IEEE Vehic. Tech. Soc. Conf., pp. 5–8, Denver, Colorado, May 10–12, 1992.

[3] T. Fujii. "Optimum Antenna Beam Tilting for Cellular Mobile Radio Systems", in Japanese, IEICE Trans. Comm. (Japan), vol. J77-B-II, no. 3, pp. 166–170, March 1994.

[4] B. Nemsic. Planung zellularer Mobilfunknetze. Dissertation, Technische Universität Wien, June 1990.

[5] D. Parsons. The Mobile Radio Propagation Channel. Halsted Press, New York, Toronto, 1992.

[6] V. Graziano. "Propagation Correlations at 900 MHz". IEEE Trans. on Vehicular Technology, vol. VT–27, no. 4, pp. 182–189, November 1978.

[7] L. F. Fenton. " The Sum of Log–Normal Probability Distributions in Scatter Transmission Systems", IRE Trans. on Communications Systems, CS–8, pp. 57–67, March 1960.

[8] M. Hata, K. Kinoshita, and K. Hirade. "Radio Link Design of Cellular Land Mobile Communication Systems", IEEE Trans. on Vehicular Technology, vol. VT–31, no. 1, pp. 25–31, February 1982.

[9] A. Papoulis. Probability, Random Variables, and Stochastic Processes. McGraw–Hill Kogakusha Ltd., Tokyo, 1965.

[10] Y. Yamada, Y. Ebine, and K. Tsunekawa. "Base and Mobile Station Antennas for Land Mobile Radio Systems", IEICE Trans., vol. E 74, no. 6, June 1991.

[11] Y. Ebine. "Antenna Characteristics in Vertically Gaussian Distributed Mobile Radio Propagation", IEEE Trans. AP–S, 83–3, pp. 1800–1803, 1990.

Performance of RSS-, SIR-based Handoff and Soft Handoff in Microcellular Environments

Per-Erik Östling

Radio Communication Systems Lab.
Dept. of Teleinformatics
Royal Inst. of Technology
S–164 40 Stockholm/Kista
SWEDEN

Abstract

In microcellular systems the "corner effect" require that handoffs can be performed fast and at high reliability. This requirement is in contrast to the desire to keep the handoff rate low, i.e., the risk for "flip-flopping" of the call between base stations should be minimized. Handoff decisions are usually based on estimates of what signal quality that can be provided by different base stations. These estimates can be constituted by, e.g., received power (RSS) and/or signal-to-interference ratio (SIR) of used and target channels. In this work we compare the performance of an RSS-based handoff with the performance of a SIR-based scheme. We also analyze the behavior of a soft handoff (SOHO) scheme where several base stations simultaneously "listen" to a mobile while only one base station is transmitting.

We show that incorporating SIR estimates of the channels in the handoff decisions can reduce the handoff rate by 50% with the same signal quality as compared to the RSS-based scheme. For the RSS-based, and the SIR-based handoff schemes, the corner effect limits the maximum allowed speed around the corner for a given signal quality due to signal outages in the uplink. The receiver diversity in the uplink for the SOHO scheme makes the uplink practically insensitive to the corner effect. For this scheme the maximum speed is limited by the quality in the downlink, which allows much higher speeds.

1 Introduction

One fundamental function in mobile cellular systems is the automatic transfer of a call from one base station to another, *handoff*, when a mobile moves between their coverage areas (cells). With the deployment of microcells in modern cellular systems, faster and more reliable handoff schemes are needed. The estimation of which base station that provides the "best" connection is usually based on measurements of *signal quality indicators* (SQI). A SQI could be constituted by, e.g., signal strength (RSS) or signal-to-interference ratio (SIR) measurements. In this treatment we compare the performance of an RSS-based handoff scheme with that of a SIR/RSS-based handoff scheme for a mobile that moves between two base stations while passing a street corner. We will also study the performance of a soft handoff (SOHO) scheme which utilizes several base stations as receivers, and thereby achieve receiver diversity to improve the signal quality in the uplink.

The radio environment in microcellular systems is characterized by the "corner effect"; a rapid drop in signal strength as a mobile turns around a corner. This means that a handoff of the call to a base station on the crossing street must be performed swiftly so that the call can be continued without interruption. Moreover, since the cell dimensions are small, a typical call will experience several handoffs and therefore each handoff must be performed very reliably as the probability of an interrupted call increases with the expected number of handoffs during a call. Because the user density will be several times higher than in macrocellular systems it is essential that a handoff scheme is designed such that decisions are distributed as much as possible.

Much work on handoff performance have used signal strength measurements of surrounding base stations to determine which base station the mobile shall be connected to [1, 2, 3, 4, 5, 6]. The focus has mainly been on the trade-off between the reaction time to changes in signal strength, and the risk for unnecessary handoffs. It was indicated in [7] that the presence of the corner effect for turning mobiles in city environments requires fast handoff decisions for sufficient handoff reliability. This is achieved at the expense of a high handoff rate for mobiles that go straight through an intersection. Lately, use of direction sensitive schemes have been suggested to reduce the probability of handoffs to base stations from which the mobile is receding [8, 9]. This diminishes the risk for "flip-flopping" of the call, and consequently reduces the handoff rate. However, some means for direction estimation is required which complicates the system.

It is not the signal strength per se that determines the signal quality; also the interference has a strong influence. The interference characteristics are dependent on the reuse pattern of channels [10]. Therefore the requirements on the handoff performance will be different for different cell plans. This was studied in [11] where it was shown that, using only a single link quality measure (RSS, uplink SIR or downlink SIR; not a combination of these) as handoff criterion does not provide optimum handoff performance. An RSS-based handoff scheme will occasionally make handoff to channels with high interference levels, and a one-way SIR-based scheme can select a channel pair with high interference level in the opposite link. With a single link SQI there is also a high risk for handoff to base stations that are not line-of-sight (LOS) with the mobile. It was therefore suggested in [11] that handoff decisions should be based on a combination of more than one SQI (RSS and SIR). The handoff performance can be further improved if *intracell* handoffs are allowed (change of channels within the same base station) when the interference suddenly increases on a channel. Such an approach has been suggested in dynamic channel allocation (DCA) systems such as the DECT system [12].

In this report we extend the work in [11] to incorporate a SIR-based scheme where the mobile measures the received signal strengths, and the downlink interference levels of available channels at surrounding base stations. Thus, it is possible to estimate the would-be downlink SIR, but not the uplink SIR, after handoff. Since the interference levels can be very different in the uplink and the downlink in city environments there is a risk for handoff to a channel pair with good signal quality in the downlink, but poor quality in the uplink. Such a situation can subsequently be resolved by an intracell handoff.

To improve the signal quality in the uplink, and therefore also reduce the intracell handoff rate, we consider a soft handoff scheme where several base stations "listen" to the user. We thus achieve receiver diversity in the uplink, and become less sensitive to the choice of channel (which is based on the downlink SIR). A similar scheme has been analyzed for a hexagonal cell system in [13] where an improved robustness against interference, and higher capacity were indicated.

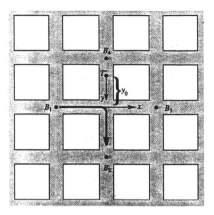

Figure 1: The mobile moves from B_1 to B_2. During the trip the mobile can be connected to any of B_1, B_2, B_3 or B_4. To model the received power along the crossing street a hypothetical transmitter T', transmitting at the received power from B_1 at the intersection, is used.

The performance of the handoff schemes will be compared in a scenario where a mobile moves between two base stations while passing a corner. The outage probability and handoff rate are estimated.

2 Handoff Schemes

Since a critical moment for radio resource management in microcellular systems is the passing of a street intersection, we will focus our attention to the performance of an *object mobile* that moves between two base stations while passing a street intersection according to Figure 1. We consider a layout where base stations are placed at every second street corner. The mobile initially connected to B_1 and moves at v m/s toward B_2. During the trip the handoff algorithm can connect the mobile to any of the four closest base stations (B_1, B_2, B_3, B_4). We will consider a typical city environment where the distance between corners is $R_c = 100$ m.

2.1 Algorithms

We will only be interested in algorithms that use SQIs that can be measured by the mobile, or by the currently used base station (e.g., the interference level, or SIR, in the uplink of other base stations can not be measured). This enables the use of distributed algorithms that involves the system in the handoff decisions to a minimum. The SQIs represent *local mean* measurements of the signal quality, that is, fluctuations due to multipath propagation are assumed to be filtered out. Moreover, the SQIs are assumed to be sampled every T_s seconds forming time discrete processes $d(n)$. The SQIs are then filtered to form the *decision processes*

$$\Delta(n) = \frac{1}{N} \sum_{k=0}^{N-1} d(n-k) \tag{1}$$

which are the actual signals used in the handoff algorithms. The SQI processes could be represented by the local mean received power $S(n)$ and/or the uplink SIR, $\Gamma^u(n) = S(n)/I^u(n)$ and/or downlink SIR, $\Gamma^d(n) = S(n)/I^d(n)$, where $I^u(n)$ and $I^d(n)$ represent the local mean interference processes of a given channel pair in the uplink and downlink, respectively. We also define min SIR of the current channel pair $\mu(n) = \min\left\{\Gamma^u(n), \Gamma^d(n)\right\}$. The algorithms may also make use of a hysteresis κ and a handoff threshold γ_h. If the SQI has been filtered according to (1) we use the bar notation ($\bar{x}$). For the handoff algorithms to attempt handoff, one or more handoff criteria must be fulfilled. The handoff criteria used here are:

$$
\mathcal{S}_i(n) = \begin{cases} 1, & \text{if } \bar{S}_i(n) > \bar{S}_0(n) + \kappa, \ \bar{S}_i(n) = \max_k\left\{\bar{S}_k(n)\right\}, \\ 0, & \text{otherwise,} \end{cases} \tag{2}
$$

$$
\mathcal{M}(n) = \begin{cases} 1, & \text{if } \bar{\mu}(n) < \gamma_h, \\ 0, & \text{otherwise,} \end{cases} \tag{3}
$$

$$
\mathcal{G}_{ij}(n) = \begin{cases} 1, & \text{if } \bar{\Gamma}^d_{ij}(n) > \gamma_h, \\ 0, & \text{otherwise.} \end{cases} \tag{4}
$$

$\mathcal{S}_i$ prevents handoff to the strongest base station B_i unless its received power at least κ dB higher than the power of the current base station S_0, $\mathcal{M}$ allows handoff if the quality of the current channel pair becomes too poor, and $\mathcal{G}_{ij}$ permits handoff to channel j at B_i only if the downlink SIR is sufficiently high. From these partial criteria the algorithm forms a *combined handoff criterion* $\mathcal{Z}(n)$ which is the product of one or more of the above criteria. The handoff algorithm can then be stated:

```
if Z(n) = 1 then
    perform handoff
else
    defer handoff
end
```

RSS-based handoff algorithm ($\mathcal{A}_1$) The simplest handoff algorithm makes a handoff to B_i if $\bar{S}_i > \bar{S}_0 + \kappa$. No attention is made to the SIRs of the current channel pair, and a target channel is selected randomly among available channels B_i. The combined handoff criterion for this scheme is written $\mathcal{Z}(n) = \mathcal{S}_i(n)$.

SIR-based handoff algorithm ($\mathcal{A}_2$) Handoff is initiated if the current channel pair has SIR less than γ_h. The handoff is directed to the strongest base station B_i if $\bar{S}_i > \bar{S}_0 + \kappa$ *and* a target channel with sufficiently high downlink SIR can be found, i.e., $\bar{\Gamma}^d_{ij} > \gamma_h$. If the SIR of the current channel pair is insufficient and the above relations don't apply, an intracell handoff should be attempted to a channel pair with uplink SIR and downlink SIR above the handoff threshold. For this algorithm the combined handoff criterion becomes $\mathcal{Z}(n) = \mathcal{S}_i(n) \cdot \mathcal{M}(n) \cdot \mathcal{G}_{ij}(n)$.

Soft handoff algorithm ($\mathcal{A}_3$) The *master base station* and channel pair is selected in the same way as for $\mathcal{A}_2$. If B_i is the master base station, transmitting on channel j, the uplink SIR of the traffic channel becomes

$$
\Gamma^u = \max_k\left\{\Gamma^u_{kj}\right\}, B_k \in \{\text{ base stations involved in soft handoff}\}). \tag{5}
$$

I.e., selection diversity is employed in the uplink. The combined handoff criterion for this algorithm is the same as for $\mathcal{A}_2$, $\mathcal{Z}(n) = \mathcal{S}_i(n) \cdot \mathcal{M}(n) \cdot \mathcal{G}_{ij}(n)$.

2.2 Performance Measures

As mentioned there are two opposing requirements to consider when designing handoff schemes: high signal quality and low handoff rate. The signal quality is represented by the minimum SIR Γ_0 of the traffic channel during the trip between B_1 and B_2. In other words, $\Gamma_0 = \min_n \{\Gamma(n)\}$ where $\Gamma(n)$ can be the SIR in either the uplink or downlink. The minimum acceptable SIR is defined by the *protection ratio* γ_0. For a protection ratio $\gamma_0 = 16$ dB we want the outage probability to be less than 1%, i.e.,

$$P_0 = \Pr\{\Gamma_0 < \gamma_0\} < 1\%. \tag{6}$$

This condition can be equivalently expressed in terms of the 1-percentile γ' of the SIR distribution. The 1-percentile is defined as

$$\gamma' : \Pr\{\Gamma_0 < \gamma'\} = 1\% \tag{7}$$

and (6) can be written $\gamma' \geq \gamma_0$.

The handoff algorithm's impact on the load on the network is indicated by the number of intercell handoffs W executed as the mobile travels between B_1 and B_2. Algorithms $\mathcal{A}_2$ and $\mathcal{A}_3$ have the possibility to perform intracell handoffs and therefore the number of intracell handoffs W' is of interest for these algorithms. The handoff management load on the network will be therefore presented in terms of the mean number of inter , and intracell handoff rates

$$\begin{aligned} \omega &= E\{W\}, \\ \omega' &= E\{W'\}. \end{aligned} \tag{8}$$

3 Models

To estimate the pathloss we make use of a model that has been suggested by Berg, et al. [14], for microcellular systems which was derived from empirical measurements at 900 MHz. Along a non-LOS (NLOS) street the received power is be modeled by a hypothetical transmitter T' located on the crossing street, using transmitter power equal to the received power at the intersection. The aggregate pathloss for a mobile that is NLOS from the base station is written as the product of the LOS pathloss between the base station and the intersection, and the pathloss along the crossing street from the hypothetical transmitter and the mobile.

To describe the pathloss model we use the co-ordinate system in Figure 1. The mobile's distance r from the base station is expressed in terms of x and y as

$$r = \begin{cases} x, & LOS, \\ R_c + y - y_0, & NLOS. \end{cases} \tag{9}$$

The base station and the hypothetical transmitter are placed at $x = 0$ and $y = 0$, respectively. Thus, T' is located a distance y_0 from the corner. The pathloss model was derived for a

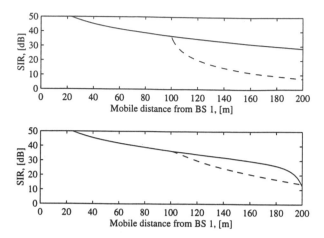

Figure 2: **Expected SIR as a function of the mobile's distance from** B_1**. Top: uplink SIR, bottom: downlink SIR. Solid line: mobile goes straight, dashed line: mobile turns.**

system where the transmitter and receiver antennas were placed 5 m and 2 m above ground, respectively. The pathloss along the LOS street is modeled as

$$L_{LOS}(x) = \left(\frac{x}{x_0}\right)^{m_1} \left[1 + \left(\frac{x}{x_L}\right)^{(m_2-m_1)q}\right]^{1/q}. \tag{10}$$

Here m_1 is the pathloss near the base station, m_2 is the pathloss after the breakpoint x_L. A bias in the pathloss can be determined by x_0 and the "smoothness" of the transition from pathloss exponent m_1 to m_2 is determined by q. Similarly, the pathloss from the hypothetical transmitter T' to the mobile is described by

$$L_{NLOS}(y) = \left(\frac{y}{y_0}\right)^{n_1} \left[1 + \left(\frac{y}{y_L}\right)^{(n_2-n_1)q}\right]^{1/q}. \tag{11}$$

Here the pathloss exponents and the breakpoint are given by n_1, n_2 and y_L, respectively. From the results in [14] we choose the following values of parameters to represent the pathloss for a typical city environment: $x_0 = 1$ m, $m_1 = 2$, $m_2 = 5$, $x_L = 200$ m, $y_L = 250$ m, $y_0 = 3.5$ m, $n_1 = 2$, $n_2 = 6$ and $q = 4$. The signal is exposed to shadow fading due to obstacles between the base station and the mobile. This fading was found to be well modeled by a multiplicative lognormal stochastic process $10^{v(r)}$. More precisely, $v(r)$ is a weakly stationary Gaussian process with zero mean and standard deviation σ_s. The autocorrelation function $\phi_{vv}(d) = E\{v(r)v(r+d)\}$ has been modeled in [15] for macrocellular systems as

$$\phi_{vv}(d) = \sigma_s^2 \rho^d. \tag{12}$$

Here ρ^d is the correlation coefficient for samples d meters apart. Pending similar models for microcellular systems we will use (12) in this environment. The correlation coefficient

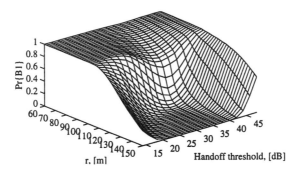

Figure 3: The probability of being connected to B_1 as a function of distance from B_1 with the handoff threshold γ_h as a parameter for A_2. Mobile's velocity $v = 1$ m/s.

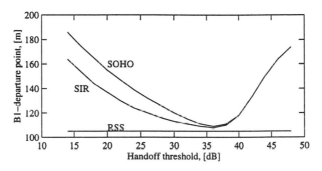

Figure 4: The position where the mobile is still connected to B_1 at 10% probability (B_1-departure point) as a function of the handoff threshold γ_h. Mobile's velocity $v = 1$ m/s.

for samples separated by 6 m was given in [14] to approximately 0.53 which corresponds to $\rho = 0.9$. The standard deviation of the fading is $\sigma_s = 3.5$ dB.

With the models described above, and assuming that the transmitter power is P_T (same power for all mobiles and base stations) and antenna gains, feeder losses, etc., can be collected in C, the received power at a mobile located r m from the base station can be written as

$$S(r) = \begin{cases} P_T C \frac{10^{v(r)/10}}{L_{LOS}(r)}, & LOS, \\ P_T C \frac{10^{v(r)/10}}{L_{LOS}(R_c) L_{NLOS}(r-R_c+y_0)}, & NLOS. \end{cases} \tag{13}$$

The aggregate interference power is modeled as the sum of the interference for all interferers using the same channel, $I = \sum_k I_k$.

Since the base stations are placed at every second intersection, each base station covers nominally one block in each direction, and channels are reused such that each base station has four closest interfering base stations at the same distance. Such a cell plan is called *full square* FS(m, n) cell plans [10], where (m, n) represent the number of streets away, horizontally and vertically, one of the nearest base stations is located. We will use the FS$(4, 2)$ in this report,

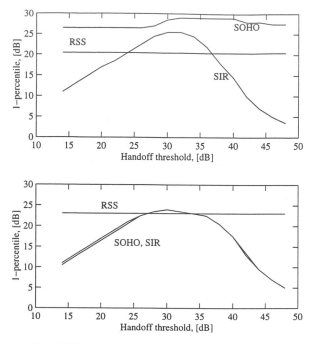

Figure 5: 1-percentiles of SIR γ' as a function of the handoff threshold γ_h. Top: uplink, bottom: downlink. Mobile's velocity $v = 1$ m/s.

and the expected the SIR in the uplink and downlink as a function of mobile position when it is connected to B_1 is shown in Figure 2.

We are interested in the relative performance of different handoff schemes, and not primarily the absolute levels of the performance measures. Therefore we will, for simplicity, assume that all channels at other base stations are occupied. It is assumed that co-channel mobiles are stationary during the object mobile's trip between B_1 and B_2. Furthermore, we assume that the actions by the object mobile will not cause the co-channel mobiles to change their channel constellations. By using such a system model the impact of new users entering the system, or co-channel users turning around the corner and becoming LOS, cannot be studied. We will assume that each base station always has $\eta = 5$ channels available.

4 Numerical Results

The numerical results were obtained by means of Monte-Carlo simulations where the mobile drove 5,000 times between B_1 and B_2 and data was collected each run. In the handoff algorithm we use $N = 2$ and a hysteresis $\kappa = 4$ dB.

Using small values of γ_h means that the mobile can move far from the corner towards B_2 before $\mathcal{M}$ becomes 1 and handoff is attempted. On the other hand, with very high

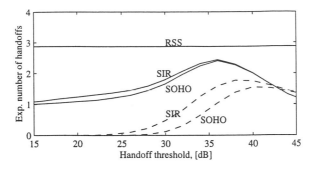

Figure 6: Expected number of handoffs as a function of the handoff threshold γ_h. Solid lines: intercell handoff rate ω, dashed lines: intracell handoff rate ω'. Mobile's velocity $v = 1$ m/s.

value of γ_h, $\mathcal{M} = 1$ long before the corner where no sufficiently strong base stations are to be found, $\mathcal{S}_i = 0$. Thus, no handoffs are likely to take place before the corner with high γ_h. After the mobile passes the intersection the target base station becomes stronger than the current. Algorithms $\mathcal{A}_2$ and $\mathcal{A}_3$ require that the downlink SIR of the target channel is higher than γ_h, and if γ_h is too high no target channels can be found ($\mathcal{G}_{ij} = 0$) until very close to the target base station. In Figure 3 this is illustrated using $\mathcal{A}_2$ for a mobile moving at $v = 1$ m/s where the probability of being connected to B_1 at distance r after the corner, $P_{B_1}(r)$, is reduced with increasing handoff threshold until $\gamma_h \approx 36$ dB after which $P_{B_1}(r)$ increases rapidly. It should be noted that 36 dB is approximately equal to the mean SIR at the corner, which means that the mobile has to turn around the corner to find a target channel with reasonably high probability. These results are condensed in Figure 4 where the B_1-departure positions r_1 are plotted (B_1-departure position, $r_1 : P_{B_1}(r_1) = 10\%$). Since $\mathcal{A}_3$ employs receiver diversity in the uplink $\mathcal{M}$ is less likely to be 1 than for $\mathcal{A}_2$ at a given distance after the corner. Thus the corresponding r_1 are further away from the corner for $\mathcal{A}_3$. For $\gamma_h > 36$ dB the r_1-values for $\mathcal{A}_2$ and $\mathcal{A}_3$ increase with γ_h since no target channels with sufficiently high downlink SIR can be found near the corner.

The handoff threshold, of course, affects the signal quality since the SIR is intimately connected to the handoff positions. To indicate the impact on signal quality, Figure 5 shows the 1-percentiles of downlink SIR and uplink SIR as a function of γ_h. As expected, the 1-percentiles for $\mathcal{A}_2$ are relatively small for small values of γ_h, but increasing with the handoff threshold since handoffs can then be made closer to the corner. Because of receiver diversity in the uplink for $\mathcal{A}_3$ the 1-percentile for the uplink SIR is fairly insensitive to the magnitude of γ_h, whereas in the downlink the 1-percentile is similar to that of $\mathcal{A}_2$. Intracell handoffs have no effect on downlink quality since all channels belonging to the same base station have the same downlink SIR. When $\gamma_h > 36$ dB the downlink signal quality is reduced for $\mathcal{A}_2$ and $\mathcal{A}_3$, whereas it is only for $\mathcal{A}_2$ the uplink quality is reduced. With magnitudes of γ_h such that the mobile remains long with B_1 (i.e., $\gamma_h < 30$ dB and $\gamma_h > 40$ dB for $\mathcal{A}_2$ and $\mathcal{A}_3$) there is a small number of handoffs as can be seen in Figure 6. The expected handoff rate, ω, increases with γ_h for $\mathcal{A}_2$ and $\mathcal{A}_3$ until $\gamma_h \approx 36$ dB where the handoff rate starts to decrease for $\mathcal{A}_2$ and $\mathcal{A}_3$ due to the lack of target channels with downlink SIR $> \gamma_h$.

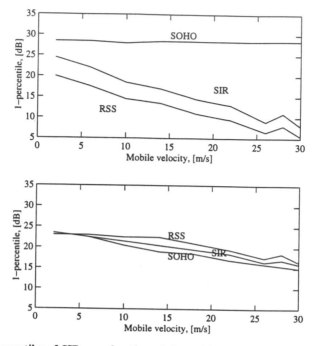

Figure 7: 1-percentiles of SIR as a function of the mobile's velocity. Top: uplink, bottom: downlink. Handoff threshold γ_h = 30 dB.

We see that the expected rate of intracell handoffs, ω', starts to become noticeable for $\gamma_h > 25 - 30$ dB. For small values of γ_h intracell handoffs are mainly due to handoffs to new base stations on channels with poor quality in the uplink, whereas for high values of γ_h intracell handoffs are made within B_1 since the the quality at B_1 decreases and no new base stations that fulfill $\mathcal{G}_{ij} = 1$ can be found.

A handoff threshold $\gamma_h = 30$ dB seems to give a good performance in terms of low handoff rate and high signal quality. With this value of γ_h (somewhat smaller than expected downlink SIR at the corner) handoffs are likely to be delayed to shortly after the corner, but not so long that the signal quality becomes poor.

In the downlink the signal qualities for all algorithms don't differ very much, Figure 7. In the uplink, Figure , however, the 1-percentile levels for $\mathcal{A}_1$ and $\mathcal{A}_2$ deteriorate fast with v whereas for $\mathcal{A}_3$ the receiver diversity makes the signal quality virtually insensitive to mobile speed. If we accept $\gamma' \geq 16$ dB in both the uplink and the downlink the maximum speed is limited by the quality in the uplink to 9 m/s and 17 m/s for $\mathcal{A}_1$ and $\mathcal{A}_2$, respectively. Algorithm $\mathcal{A}_3$, on the other hand is limited by the quality in the downlink, and and the mobile can be allowed to drive at 25 m/s.

5 Conclusion

One important issue in city cellular systems is the handoff performance for a mobile that makes a turn at an intersection. We have compared the performance of an RSS-based handoff scheme, a SIR-based handoff scheme and a soft handoff scheme in terms of the trade-off between handoff reliability and handoff rate.

We found that with a proper choice of the handoff threshold for the SIR-based scheme, the handoff rate can be reduced by 50% and still achieve higher signal quality than if only signal strength is is the handoff criterion. The magnitude of the handoff threshold is important. With a too low γ_h the mobile will not try to make a handoff until deep inside the target cell, and with a too high γ_h no target channels with sufficiently high downlink SIR can be found until very close to the target base station. A good compromise is to choose γ_h somewhat smaller than the expected downlink SIR at the corner.

The uplink is very sensitive to the corner effect. For RSS-based scheme and the SIR based scheme the maximum velocity for 1% outage probability in either the uplink or downlink was limited by the quality in the uplink to approximately 9 m/s and 17 m/s, respectively. Using receiver diversity in the uplink (soft handoff) makes the uplink virtually insensitive to the street corner effect and the maximum velocity is restricted by the downlink quality to approximately 25 m/s.

We have thus showed that using a handoff threshold can greatly enhance handoff performance in microcellular systems, and that the corner effect can be disregarded if soft handoff with uplink receiver diversity is employed.

References

[1] R. Vijayan and J. Holtzman, "Model for analyzing handoff algorithms," *IEEE Transactions of Vehicular Technology*, vol. Vol. 42, Aug 1993.

[2] M. D. Austin and G. L. Stüber, "Analysis of a window averaging hand-off algorithm for microcellular systems," in *Proceedings: 42nd IEEE Vehicular Technology Conference*, (Denver, Co), May 1992.

[3] M. Gudmundson, "Analysis of handover algorithms," in *Proceedings: 41st IEEE Vehicular Technology Conference*, (St. Lois, Mo), 1991.

[4] S. T. S. Chia, "The control of handover initiation in microcells," in *Proceedings: 41st IEEE Vehicular Technology Conference*, pp. 531–536, 1991.

[5] A. Murase, I. C. Symington, and E. Green, "Handover criterion for macro and microcellular systems," in *Proceedings: 41st IEEE Vehicular Technology Conference*, (St. Lois, Mo), 1991.

[6] S. Tekinay and B. Jabbari, "Handover and channel assignment in mobile cellular networks," *IEEE Communications Magazine*, pp. 42–46, November 1991.

[7] O. Grimlund and B. Gudmundson, "Handoff strategies in microcellular systems," in *Proceedings: 41st IEEE Vehicular Technology Conference*, (St. Lois, Mo), 1991.

[8] M. D. Austin and G. L. Stüber, "Velocity adaptive handoff algorithms for microcellular systems," *IEEE Transactions of Vehicular Technology*, vol. 43, pp. 549–561, August 1994.

[9] A. Sampath and J. Holtzman, "Estimation of maximum doppler frequency for handoff decisions," in *Proceedings: 43rd IEEE Vehicular Technology Conference*, (Secaucus, NJ,), pp. 859–862, 1993.

[10] M. Gudmundson, "Cell planning in manhattan environments," in *Proceedings: 42nd IEEE Vehicular Technology Conference*, (Denver, Co.), May 10–13, 1992.

[11] P.-E. Östling, "Handoff performance in city environments using single link handoff criteria," Tech. Rep. TRITA–IT–R 94-33, Royal Inst. of Technology, January 1995.

[12] H. Eriksson and R. Bownds, "Performance of dynamic channel allocation in the DECT system," in *Proceedings: 41st IEEE Vehicular Technology Conference*, (St. Lois, Mo.), May 19–22 1991.

[13] W. C. Y. Lee, "Smaller cells for greater performance," *IEEE Communications Magazine*, Nov 1991.

[14] J.-E. Berg, R. Bownds, and F. Lotse, "Path loss and fading models for microcells at 900 MHz," in *Proceedings: 42nd IEEE Vehicular Technology Conference*, (Denver, Co.), May 10–13, 1992.

[15] M. Gudmundson, "Correlation model for shadow fading in mobile radio systems," *Electronic Letters*, vol. Vol. 27, Nov 1991.

On the Use of Signal-to-Noise Ratio Estimation for Establishing Hidden Markov Models

William H. Tranter and Otto H. Lee
Department of Electrical Engineering
University of Missouri-Rolla
Rolla, MO 65401

Abstract

A technique for developing a two-level finite-state Markov model that can accurately describe a frequency nonselective slowly fading channel is illustrated. A dynamic channel model is first found by monitoring the receiver input signal-to-noise ratio (SNR). Each channel state is then embedded with a binary error Fritchman model. From the Markov model, system performance parameters, such as the bit-error-rate (BER), can be determined directly.

1. Introduction

In Monte-Carlo simulations of mobile communication systems operating in a fading environment, a large number of error events are needed in order to provide a reliable estimate of the error probability at a particular SNR at the receiver input. This in turn requires considerable computation time, even for systems with moderate complexity.

For BER estimation, it is usually sufficient to model the channel statistics without modeling the details of the underlying physical phenomena which cause the errors to occur. Thus, an alternate simulation approach is to characterize the sequence of channel events by an appropriate statistical model. In a mobile communications environment, the received signal typically suffers from multipath propagation effects that result in signal fading. Modeling of such a channel in detail is a formidable task. It has been pointed out that if environmental attributes are slowly varying, then a real land mobile satellite channel can be viewed as quasistationary [1]. That is, the channel at any instant of time can be seen as operating at one of a finite number of channel states. This paper investigates a two-level Markov model that is capable of modeling the channel dynamics as well as the error statistics.

2. Hidden Markov Model

Consider an n-state Markov chain with states s_i, $i = 1, 2, \cdots, n,$, and a set of M possible events E_i, $i = 1, 2, \cdots, M$. Assuming that the Markov chain is in some initial state, and one of the M events is generated by a corresponding probabilistic function. At each integer multiple of a clock period T, a transition is made from one state to another state according to the state transition probability matrix, and a new event outcome is generated. Since each probabilistic function is capable of producing any of the M event outcomes, the same observed event can be produced by more than one state. Thus, the states of the underlying Markov chain cannot be observed directly from the sequence of event outcomes; hence the name hidden Markov model. The conditional probability $\Pr(\text{next state} = s_j \mid \text{present state} = s_i) = a_{ij}$, $i, j = 1, \cdots, n$, is called the transition probability from state i to state j. The set of probabilities $\Pr(\text{present state} = s_i) = p_i$, $i = 1, \cdots, n$, is called the initial state probabilities and can be expressed as a $1 \times n$ row vector $\mathbf{p} = [p_1 \; p_2 \; \cdots \; p_n]$. The Markov chain is said to be stationary if $\mathbf{p} = \pi$, such that $\pi \mathbf{A} = \pi$ and $\pi \mathbf{1} = 1$, where $\mathbf{A}$ is the transition probability matrix, and $\mathbf{1}$ is a column vector of n ones.

Let 0 and 1 represent a correct and an erroneous reception, respectively. Let 0^n represent n consecutive correct receptions. The probability of an n symbol error-free run is the conditional probability that given an error has occurred, it will be followed by n or more error-free symbols. That is,

$$\Pr\{n \text{ symbol error-free run}\} = \Pr\{0^n \mid 1\} \tag{1}$$

and

$$\sum_{n=1}^{\infty} \Pr\{0^n \mid 1\} = 1 \tag{2}$$

The error-free run distribution is represented by a plot of $\Pr\{0^n \mid 1\}$ versus the length n.

A simple-partitioned Fritchman model [2] is a finite state Markov model where there are no transitions between the error-free states. This allows multiple degrees of memory to exist in the model, and therefore it is suitable for modeling real communication channels. Furthermore, there is only one single-error state, and therefore the conditional probability of error given that the process is in an error-free state is zero, and the conditional probability of error given that the process is in an error state is one. This allows the error-free run distribution to uniquely specify the single-error state model. In other words, the model parameters can be found from the error-free run distributions, and vice versa.

3. Fading Channel Model

The waveform-level simulation serves to provide raw data for Markov modeling without resorting to actual field measurements. Figure 1 illustrates a block diagram of a digital communications link with a Rician fading channel. In a mobile communications environment, the channel statistics vary from one distribution to another depending on the topography. The fading channel model considered here consists of two channel models. A fading channel model for urban areas and a fading channel model for open areas will be described. These channel models are taken from existing results in the area of mobile communications [1], [3], [4]. In both cases, a narrowband signal with frequency nonselective fading is assumed so that signal fading can be treated as multiplicative noise. Another assumption is that the received signal undergoes slow fading, such that the signal envelope is constant over a symbol period.

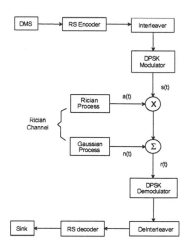

Figure 1. Block diagram of a communications link with a Rician fading channel.

3.1 Fading Channel Model For Urban Areas

When a mobile receiver is traveling in an urban area, the direct line-of-sight path is often completely obstructed by high-rise buildings. The received signal, $r(t)$, can be characterized as a sum of reflected signals, each having an unknown attenuation, time-delay and phase-shift plus an additive white Gaussian noise (AWGN) component, $n(t)$. Thus

$$r(t) = \sum_{i=1}^{N} d_i(t) + n(t) \tag{3}$$

The instantaneous value of the ith reflected signal can be written as $d_i(t) = D_i(t) \cos[2\pi f_c t + \phi_{di}(t)]$, where $D_i(t)$ is the signal envelope, and f_c is the carrier frequency. In addition $\phi_{di}(t) = 2\pi f_{di} t + \phi_i$ is the received signal phase deviation, where ϕ_i is the transmitted signal phase, and f_{di} is the Doppler frequency shift. For N sufficiently large, the sum of the reflected signals has an envelope which has a Rayleigh distribution [4]

$$f(D) = \frac{D}{\sigma_d^2} \exp\left(\frac{-D^2}{2\sigma_d^2}\right) u(D) \tag{4}$$

where $u(D)$ is the unit step function. The phase is uniformly distributed in $(-\pi, \pi)$. Since the noise component is a zero mean uncorrelated complex Gaussian process, the envelope of the received signal $r(t)$, as defined by (3), also has a Rayleigh distribution.

3.2 Fading Channel Model For Open Areas

When the mobile receiver is traveling in an open area, the direct line-of-sight signal can often be received without any obstructions. Thus the received signal can be characterized as a sum of a direct line-of-sight signal, $s(t)$, a road reflected diffused signal, $d(t)$, and an AWGN component, $n(t)$. This can be written as $r(t) = s(t) + d(t) + n(t)$. The line-of-sight transmitted signal $s(t)$ is assumed to have a constant envelope. The diffused component $d(t)$ has a Rayleigh distributed envelope as given in (4), and the received phase is uniformly distributed from zero to 2π. The sum of a constant direct path envelope and a Rayleigh distributed diffused path envelope results in a signal envelope $R(t)$ with a Rician distribution [4]

$$f(R) = 2R(K+1)e^{-R^2(K+1)-K}I_o(2R\sqrt{K(K+1)})$$
(5)

where I_o is the modified Bessel function of order zero, and K is the direct-to-multipath signal power ratio, or the Rician factor. By varying K, a family of Rician distributions corresponding to different fading conditions can be obtained. Figure 2 shows a family of Rician probability density functions with different values of the Rician factor. Based on these fading channel models, a waveform-level simulation of the Rician process is generated from the method given by Lutz and Plochinger [5].

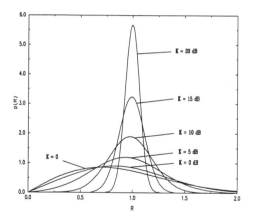

Figure 2. Rician probability density functions with different values of Rician factors.

3.3 Channel Dynamics

Most existing simulation methods simulate the fading channel dynamics by switching between Rician fading, with a fixed Rician factor for good channel conditions, and a Rayleigh fading for bad channel conditions. Such a switching scheme is easy to implement, but lacks flexibility. Within a homogeneous geographical area, a Rician channel with a fixed Rician factor gives a valid description of a frequency nonselective slowly fading condition. However, for less well-defined geographical boundaries, the simulation should take into account the changes in the channel conditions.

The approach here adapts to multiple fading conditions by allowing the Rician factor to change during the simulation. After the baseline Rician process has been established, a Markov process, with each state representing a different fading condition, or Rician factor, is generated. The first step is to decide the number of states (the number of Rician factors allowed), and the transition probability matrix of the Markov chain. Then, according to the chosen parameters, a sequence of Markov states can be generated and stored into a data file. The next step is state assignment. A distinct Rician factor is assigned for each Markov state. The length of each state, in terms of the number of binary symbols, is specified before the simulation begins. During the simulation, a counter is used to keep track of the state duration. Once a new Markov state is entered, the simulator reads the new state number and the corresponding Rician factor which will be used to determine the parameters necessary for simulating the new fading condition.

4. Markov Modeling Techniques

The process of modeling the channel dynamics and the error statistics consists of two phases. The first phase involves applying an SNR estimation technique to estimate the Rician factor of a frequency nonselective slowly fading channel as described in Section 3. Since the Rician factor serves as an indication of the depth of the fades, monitoring the changes in the Rician factors, and their transitions over a period of time, allows a Markov model that depicts the channel dynamics (*the dynamic model*), to be obtained. In the second phase, a binary simple-partitioned Fritchman model is found from the experimental data using exponential curve fitting. The entire channel model can be viewed as having a different Fritchman model embedded into each of the dynamic model states.

4.1 Fritchman Model Estimation [2]

Recall that the error-free run distribution uniquely specifies a simple partitioned Fritchman model. In fact, the error-free run distribution can be written in terms of the transition probabilities as,

164

$P(0\,^m|\,1) = \sum_{i=1}^{N-1} a_{Ni} a_{ii}^{m-1}$, $m \ge 1$. Thus, the transition probability matrix of the model can be determined by first plotting the error-free run distribution of the experimental data. Then for an N-state model, the distribution is used to fit a sum of $N-1$ exponential functions of the form

$$P(0^m|\,1) \cong \sum_{i=1}^{N-1} B_i e^{\,b_i m} \qquad (6)$$

By matching the coefficients and the exponents in (6) with the transition probabilities we have $p_{ii} = e^{\,b_i}$, $i = 1, \cdots, N-1$ and $p_{Ni} = B_i e^{\,b_i}$, $i = 1, \cdots, N-1$. The transition probabilities p_{iN}, $i = 1, \cdots, N$ can be obtained from the fact that $\sum_{j=1}^{N} p_{ij} = 1$, $i = 1, \cdots, N$.

4.2 SNR Estimation and Geometric Interpretation in Hilbert Space [6]

For a linear time-invariant distortionless system, the signal $z(t)$ at any point in the system is an amplitude-scaled and time-delayed version of the input reference signal $x(t)$. Therefore, we can write

$$z(t) = Ax(t - \tau) \qquad (7)$$

where A is the system gain, and τ is the group delay from the system input to the point in the system at which the SNR is to be estimated. Let $x(t)$ be the reference signal with power P_x, and $y(t)$ be the measurement signal with power P_y, such that

$$y(t) = z(t) + n(t) + d(t) \qquad (8)$$

where $n(t)$ is the external additive noise, and $d(t)$ is the signal dependent distortion induced by the system. For such a system, the SNR of the measurement signal is found to be [6,7]

$$SNR = \frac{R_m^2}{P_x P_y - R_m^2} = \frac{\rho^2}{1 - \rho^2} \qquad (9)$$

Since we are dealing with physical signals having finite energy, the set of signals $x(t)$, $y(t)$, $z(t)$, $d(t)$, and $n(t)$ belong to the Hilbert space L^2 $[t_1, t_2]$, where $[t_1, t_2]$ is the closed finite observation interval. Let $\mathcal{Y}$ be the linear vector space defined over the set of all possible $z(t)$, $d(t)$, and $n(t)$. Also, let δ be a closed linear subspace of $\mathcal{Y}$ defined over the set of possible $z(t)$ and $d(t)$. From the projection theorem, for every function $\mathbf{y} \in \mathcal{Y}$, there exists a function $\mathbf{s} \in \delta$, called the projection of $\mathbf{y}$ on δ, which satisfies the condition $\|\mathbf{y} - \mathbf{s}, \mathbf{s}\| = 0$ or $\|\mathbf{n}, \mathbf{s}\| = 0$, where $\|\mathbf{f}_1, \mathbf{f}_2\| \equiv \left(\int_{t_1}^{t_2} f_1(t) f_2(t) dt \right)^{1/2}$ is the inner product defined over L^2 $[t_1, t_2]$, or simply the L-2 norm. Thus, the noise component is orthogonal to the sum of the envelope distortionless system output plus the system distortion.

We further let $\mathcal{J}$ be a closed linear subspace of δ which is defined over the set of all possible $z(t)$, the amplitude-scaled and time-shifted version of the reference signal $x(t)$. By invoking the projection theorem again in the subspace δ, for every function $s \in \delta$, there exists a function $z \in \mathcal{G}$, the projection of s on $\mathcal{G}$, which is orthogonal to the system distortion $\mathbf{d}$. Figure 3 depicts the orthogonal decomposition of the system output in $\mathcal{Y}$ onto the subspace δ and $\mathcal{G}$. From Figure 3, it can be seen that z is also the direct orthogonal projection of y onto $\mathcal{G}$ with ε as the resultant of $\mathbf{n}$ and $\mathbf{d}$. Thus the mean-squared error, or the total noise power, is minimized, and the geometric interpretation of the SNR is

$$SNR = \frac{\| z \|^2}{\| \varepsilon \|^2} \qquad (10)$$

which agrees with our previous results.

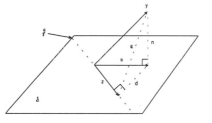

Figure 3. Geometric interpretation of signals $y = z + d + n$ in Hilbert space. δ is the subset of all possible outputs in the absence of random noise, and $\mathcal{G}$ contains the set of all possible time-translation of the reference signal.

For the moment, assume that the system under consideration is distortionless such that $\mathbf{d} \equiv 0$. The slowly fading channel, as described in Section 3, introduces multiplicative fading onto $z(t)$. Since the subspace δ contains the set of all possible outputs in the absence of random noise, it contains the direct path signal z plus the effect of multipath fading $\mathbf{f}$. From the projection theorem, the direct path signal is the orthogonal projection of the received fading signal y onto the subspace $\mathcal{G}$. Thus with $\mathbf{d}$ replaced by $\mathbf{f}$, Figure 3 can be used as a geometric interpretation of signals in a fading channel in the context of Hilbert space. We now want to estimate the direct-to-multipath signal power ratio $\hat{K} = \| z \|^2 / \| f \|^2$. Since $\| f \|^2$ cannot be estimated directly by using the receiver output as the measured signal, we estimate the Rician factor as

$$\hat{K} = \frac{\| z \|^2}{\| \varepsilon \|^2 - \| n \|^2} \qquad (11)$$

First we obtain an estimate of $\| n \|^2$ without the effects of fading. This can easily be accomplished in a simulation environment by executing a part of the simulation with the fading turned off. In practice, this can also be accomplished for a certain kind of fading channel. A land mobile satellite communication channel where only the receiver is in motion is an example. The noise component present in the downlink can be used as an approximation of $\mathbf{n}$. By defining the SNR in the fading channel as

$$S_f = \frac{\|\mathbf{z}\|^2}{\|\mathbf{n}\|^2} \tag{12}$$

and the noisy estimate of the Rician factor as

$$K_N = \frac{\|\mathbf{z}\|^2}{\|\mathbf{f}+\mathbf{n}\|^2} \tag{13}$$

an estimate of $\hat{K}$ can be obtained by expressing (11) in terms of (12) and (13) as

$$\hat{K} = \frac{S_f K_N}{S_f - K_N} \tag{14}$$

4.3 Model Parameter Estimation

During the simulation an error sequence is obtained by comparing the source sequence and the received sequence. The error sequence is then segmented into 4096-sample blocks, and a Rician factor estimate is calculated over each data block until the error sequence is exhausted. The size of the Markov model is usually difficult to estimate, and usually requires some trial and error. In this case, a histogram of the Rician factor estimates from 0.5dB to 20.5dB with a bin size of 1dB is obtained. As shown in Figure 2, for $\hat{K} < 0$dB, the Rician pdf approaches the Rayleigh pdf. Therefore, separate bins for estimates with $\hat{K} < 0$dB become unnecessary, and all estimates with $\hat{K} \leq 0.5$dB are put into one separate bin. For estimates with $\hat{K} > 20$dB, the Rician pdf approaches the Gaussian pdf, and therefore estimates with $\hat{K} > 20.5$dB are also put into a separate bin. It is reasonable to assume that over a long observation period, estimates with low relative frequency are due to a transition occurring within a data block, and estimates with a high relative frequency are a result of the channel remaining in a state that characterizes the physical environment for a longer period of time. Thus the number of states is estimated as the number of peaks in the histogram. After determining the size of the model, N, the following parameters are determined:

n_b -- the total number of data blocks,

n_i -- the total number of data blocks belonging to state i, $i = 1, \cdots, N$,

n_{ij} -- the total number of transitions between each pair of states, s_i and s_j, $i, j = 1, \cdots, N$, and

t_i -- the total number of transitions out of state i, $i = 1, \cdots, N$.

The state probabilities can then be calculated as $p_i = n_i/n_b$, $i = 1, \cdots, N$, and the transition probabilities can be calculated as $a_{ij} = n_{ij}/t_i$, $i, j = 1, \cdots, N$, respectively.

5. Results

Data obtained from a simulated fading channel, which switches among a Rayleigh state, a Rician state with $K_1 = 8$dB, and a Rician state with $K_2 = 15$dB, is analyzed. These states are chosen such that significantly different fading environments are included in the simulation. As a starting point a Markov channel model is chosen having a transition probability matrix

$$A = \begin{bmatrix} 0.795 & 0.200 & 0.005 \\ 0.200 & 0.600 & 0.200 \\ 0.010 & 0.150 & 0.750 \end{bmatrix} \tag{15}$$

and the corresponding state probabilities

$$\pi = \begin{bmatrix} 0.3792 & 0.3229 & 0.2979 \end{bmatrix} \tag{16}$$

The diagonal elements of the transition probabilities are chosen such that upon entering a new state, the channel has a high probability of staying in the same state before switching to another state.

Each Monte-Carlo simulation employs 10^7 samples. A Rician factor estimate is obtained after each 4096-sample block is elapsed. This gives a total of 2441 Rician factor estimates. This is repeated for SNRs from 0dB to 50 dB in 5dB increments as measured at the receiver. Rician factor estimates are quantized from 0.5dB to 20.5 dB. Rician factor estimates of 0.5dB or less are grouped together as 0dB, and Rician factor estimates of 20.5dB or more are also grouped together as 21 dB. Thus a discrete function showing the frequency of occurrence of each discrete Rician factor estimate can be obtained. Furthermore, a separate discrete function can be found at each different SNR. A three-dimensional plot of the set of discrete functions is given in Figure 4. The y-axis represents the discrete Rician factor estimate from 0dB to 21 dB in 1dB increments. Each point on the surface of the three-dimensional plot represents the frequency of occurrence of the discrete Rician factor estimate at a particular SNR.

From Figure 4 it can be seen that at low SNR the channel states are masked by the background noise, and therefore cannot be observed. In fact, at an SNR of 0dB, the results tell us that the channel consists of only one state -- a very noisy Rayleigh state. As the channel SNR increases, the hidden Markov states gradually reveal themselves. Basically, once the SNR is higher than the highest Rician factor in the channel, the underlying Markov states can be determined by locating the maxima of the three-dimensional plot. By looking at the frequency of occurrence, three distinct states are observed. These are a Rayleigh state, a Rician state with $K_1 = 8$dB, and a Rician state with $K_2 = 15$dB. In this case, all three Markov states have been identified correctly. Their corresponding frequencies of occurrence are 954, 776, and 711, respectively. The state probabilities are estimated from their relative frequency as

$$\tilde{\pi} = \begin{bmatrix} 0.3908 & 0.3179 & 0.2913 \end{bmatrix} \tag{17}$$

To estimate the transition probabilities, the transition from one Rician factor estimate to another is recorded during the simulation. A three dimensional plot of the number of occurrences of the transitions is

168

given in Figure 5. The height of each point on the surface gives the information on how often a transition has been taken place. Since the hidden Markov states have been identified, transitions among the 21 states are simplified to only include the transitions among the three identified Markov states. This is done by finding the midpoint between each pair of adjacent Markov states as the state boundary. Adding all the transitions within the state boundaries of each Markov state, a simplified 3-state transition matrix can be found by normalizing the number of transitions among states by the total number of transitions. This yields an estimate of the transition probability matrix

$$\tilde{A} = \begin{bmatrix} 0.768258 & 0.207865 & 0.023877 \\ 0.189433 & 0.582474 & 0.228094 \\ 0.018008 & 0.184322 & 0.797670 \end{bmatrix} \tag{18}$$

By comparing (18) and (17) with (15) and (16), it can be seen that results produced by the dynamic channel model estimation procedure are close to the actual values. In the next section, more meaningful comparisons in terms of BER performance will be given.

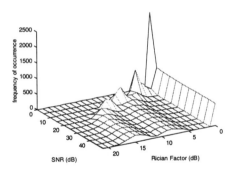

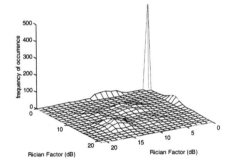

Figure 4. Frequency of occurrence of different Rician factor estimates over a range of signal-to-noise ratios for the test channel.

Figure 5. Transition activities among the three dynamic channel model states.

After identifying the dynamic model, each 4096-sample block obtained from the simulated fading channel are grouped together according to the dynamic channel state in which they belong. This separates the entire output sample sequence into three groups, and from that, three error-free run distributions can be obtained. This is repeated for SNRs from 0dB to 50 dB in 5dB increments. Exponential curve fitting is then carried out for each of the error-free run distributions with a mean-squared error of no more than 10^{-6} . With the exponential fit of each error-free run distribution, a simple-partitioned Fritchman model can be obtained by the method as given in Section 4.1. This gives a simple-partitioned Fritchman model embedded in each of the dynamic channel model state at a particular SNR. With the two-level Markov channel model, system performance can be determined directly. Figure 6 shows the uncoded BER of each individual dynamic model state, and Figure 7 shows the uncoded BER obtained from 2-level Markov modeling and from a Monte-Carlo waveform-level simulation. It can be seen that the Markov modeling results agree well with the Monte-Carlo simulation results.

The condition of each dynamic channel state can be observed through the set of BER curves. From Figure 6, the results obtained from the dynamic model estimation alone show that the three distinct hidden Markov states are not observable until the SNR reaches 35dB. The BER curves obtained from the Markov model, which starts to diverge after the SNR has reached 25dB, agree with the above results. For SNRs beyond 25dB, different degrees of fading can be observed. It is interesting to see that even though the channel has undergone different degrees of fading, the average BER is close to the BER of a Rician channel with a Rician factor of 8dB. Thus a single estimate of the average BER alone fails to provide information concerning the channel dynamics that can be obtained through Markov modeling.

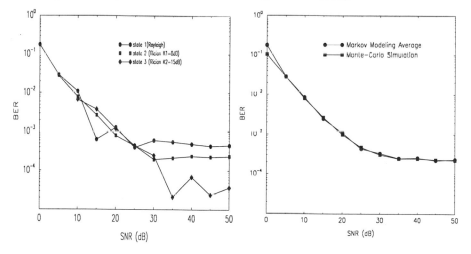

Figure 6. Uncoded BER of each dynamic model state of a Markov modeled fading channel.

Figure 7. Uncoded BER obtained from Markov modeling and Monte-Carlo simulation.

6. Conclusions

A two-level Markov modeling method is carried out. The first step of the modeling procedure utilizes a classical SNR estimation method to estimate the fading channel dynamics. The transitions among different fading conditions are determined by monitoring the changes in the Rician factor estimates. When the system SNR is low, the hidden Markov states are masked by the background noise, and become unobservable. When the SNR exceeds the largest Rician factor of the fading channel, the estimation results accurately determines the three hidden Markov states, and the resulting Markov model is found to be very close to the true model. This demonstrates the usefulness of the SNR estimation method for establishing dynamic models for mobile fading channels. The second step of the modeling procedure involves estimating a binary simple-partitioned Fritchman model from the error sequence. Exponential curve fitting is performed on the error-free run distributions, and binary Markov models are constructed from the parameters. The overall result is a two-level Markov model with a binary Fritchman model embedded into each of the dynamic model states.

The BER derived from the Markov model comes remarkably closed to the one obtained from a Monte-Carlo simulation. This paper has demonstrated that multilevel Markov modeling provides a feasible solution to accurately model complicated channels.

References

[1]   Vucetic, B. and J. Du, "Channel Modeling and Simulation in Satellite Mobile Communication
       Systems," *IEEE Journal on Selected Areas in Communications*, Vol. 10, No. 8, October 1992, pp.
       1209-1218.

[2]   Fritchman, B. D., "A Binary Channel Characterization Using Partitioned Markov Chains," *IEEE
       Transactions on Information Theory*, Vol. IT-13, No. 2, April 1967, pp. 221-227.

[3]   Lutz, E., et al., "The Land Mobile Satellite Communication Channel -- Recording, Statistics, and
       Channel Model, " *IEEE Transactions on Vehicular Technology*, Vol. 40, No. 2, May 1991,
       pp373-385.

[4]   Jakes, W. C. Jr., Microwave Mobile Communications, John Wiley and Sons, New York, 1974.

[5]   Lutz, E. and E. Plochinger, "Generating Rice Processes with Given Spectral Properties," *IEEE
       Transactions on Vehicular Technology*, Vol. VT-34, No. 4, November 1985, pp. 178-181.

[6]   Jeruchim, M. C. and R. J. Wolfe, "Estimation of the Signal-to-Noise Ratio (SNR) in Communication
       Simulation," *IEEE GLOBECOM Conference*,1989, pp. 1274-1278.

[7]   Lee, H. O., Modeling and Performance Evaluation of Multipath Fading Channels by Hidden Markov
       Model, Ph.D. Dissertation, University of Missouri-Rolla, Missouri, 1995.

On the Performance of 4-Phase Sequences in Asynchronous CDMA Systems

Mark D. Burroughs
Stephen G. Wilson

Department of Electrical Engineering
University of Virginia
Charlottesville, Virginia

December 1, 1994

Abstract- *CDMA systems use code sequences to separate users. At present 2-phase sequences are primarily used as spreading sequences for these systems. Boztas et al [1] have developed 4-phase sequences with maximum periodic cross-correlation values that are smaller than those for 2-phase sequences and that asymptotically approach lower bounds on maximum cross-correlation. While a probabilistic analysis indicates that the performance of both types of sequences is equal, a deterministic analysis shows that 4-phase sequences indeed offer better average probability of error performance than 2-phase sequences for cases analyzed thus far.*

I. An Introduction to CDMA Code Sequences

In code division multiple access (CDMA) systems, users are separated not by time or frequency but by code sequences assigned to each user-receiver pair. Each user's data is multiplied by a user-specific code before transmission. In a multiple access system, the receiver correlates code modulated signals from many users with the code of the user it wants to detect. High autocorrelation values of the sequences at zero time shift compared to other time shifts help the correlation receiver to synchronize with the transmitter. Low cross-correlation values between sequences at all time shifts decrease other user interference, allowing the correlation receiver to better synchronize with the user and to determine what data the user sent.

Thus, the periodic autocorrelation and cross-correlation values of CDMA code sequences greatly influence the performance of the system in which they are used. For cellular CDMA systems with large distances between fast moving mobiles the cross-correlation values of codes are especially important because there is no time synchronization of mobile transmissions. (If synchronization was used in the CDMA system Hadamard sequences could be used due to their zero cross-correlation value at zero time shift between sequences. [2] Actually, this is what Qualcomm uses in the forward link of their CDMA system because the base station transmits to all of the mobiles at once so to the mobile all return signals are synchronized. [3]) Good code sets have low cross-correlation values for any time shift between codes and a high autocorrelation value for zero time shift. Welch's [4] bound places a lower bound on the maximum cross-correlation (C_{max}) for code sets based on the number of codes M and the uniform length L of the sequences. This lower bound along with bounds by Sidelnikov [5] were evaluated by Kumar and Moreno [6] and generalized to be (normalized to the autocorrelation L at zero time shift):

$$C_{max} = O(\sqrt{2L}), \quad q = 2 \ (2-phase)$$
$$C_{max} = O(\sqrt{L}), \quad q > 2 \ (3-phase, 4-phase, etc.)$$

II. Gold and Family A Code Sets

Most CDMA systems today use 2-phase Gold sets for signature sequences. These Gold sets are created by the modulo-two ($\oplus$) addition of two length L m-sequences [7], sequences that have autocorrelation values of L (for zero time shift) and -1 (for all other time shifts). For two length L sequences, the Gold set includes the two original sequences and all of the modulo-two additions of one sequence and the other sequence cyclically shifted in time, for a total of $M = L + 2$ codes per sequence set [7]. These Gold codes are asymptotically optimal with respect to the given bound on C_{max} for $q = 2$.

That the C_{max} value for sets with $q > 2$ is $\sqrt{2}$ lower than that for $q = 2$ caused researchers to look for code sets with $q > 2$ that would have smaller cross-correlation values and possibly better performance than the Gold sets. In 1992 Boztas et al [1] reported the discovery of a 4-phase code set (which they called Family A) that asymptotically achieves the given bound on C_{max} as L goes to infinity. The Family A set was found after an investigation of sequences over Z_4 generated by irreducible polynominals belonging to $Z_4[x]$. Using these polynominals the Family A code set can be generated with a shift register circuit and has C_{max} values of $C_{max} \leq (1+\sqrt{L+1})$. The uniform length of these sequences is L and the number of sequences in the set is $M = L + 2$. Comparing this 4-phase set to the Gold set, both sets have equal length and number of sequences but the 4-phase set has a lower C_{max}.

III. Cross-Correlation Plots

It is important to remember that the given bounds on C_{max} only consider periodic cross-correlation values between code sequences. In practical systems, there is data modulation (bit-switching) which causes parts of sequences to be negated. The question of whether data modulation greatly affects the cross-correlation properties of 2-phase or 4-phase sequences in asychronous systems was investigated. A MATLAB program was written which counted cross-correlation values for 2 and 4-phase sets with and without data modulation for sequence lengths $L = 7$ and 15. The output of the program, plotted in Figure 1, shows the real part of the cross-correlation values in histogram form. (The 4-phase code sets generate both real and imaginary values and the effect of this is discussed in the next section of this paper.) A list of Family A generator polynomials can be found in [1] and the code sets used to generate Figure 1 are listed in Appendix I. Note that for the 4-phase sequences the mapping is from $(0, 1, 2, 3)$ to $(1, j, -1, -j)$.

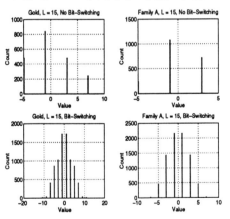

Figure 1 (Carrier phase of $0°$ assumed)

When data modulation (bit-switching) was not used, it was found (as expected) that the Family A periodic cross-correlation values matched those predicted in [1] and that the maximum periodic cross-correlation value for the Gold set was greater than the given lower bound on C_{max} for 2-phase sequences. The histograms illustrate general trends found during this research: data modulation leaves a symmetric p.d.f. for multi-user interference and increases the worst-case correlation values somewhat. With bit-switching the C_{max} values for the 4-phase case were still lower than the C_{max} values for the 2-phase case. Using MATLAB to find the variance of these distributions gives the following table:

Variance Values (L = 15)

Code Set	Variance
Gold, No Bit–Switching	14.18
Family A, No Bit–Switching	6.64
Gold, Bit–Switching	14.18
Family A, Bit–Switching	6.79

Table 1 (Carrier phase of 0 assumed)

The table values indicate that the variance of the multi-user interference is virtually unchanged by data modulation. This result and the smaller correlation values for the 4-phase codes with data modulation points to a superior bit error probability performance in asynchronous CDMA systems.

IV. The Effect of Phase Differences on Correlation Values

In the last section histograms were used to illustrate how periodic cross-correlation values between sequences were affected by data modulation in a practical system. Phase differences also affect cross-correlation values because the receiver only sees the real part of these correlation values. The distribution of the cross-correlation values for 2-phase Gold and 4-phase Family A code sets with L = 15, bit-switching, and M = 17 sequences is shown in Figure 2. It is easiest to think of phase differences as altering the mapping of cross-correlation values to the real axis in that phase differences change the orientation of the real axis. A 45° phase difference between users would move the real axis from its horizontal position to a diagonal position and correlation values would map to this new axis. When the phase difference between users is 90° the correlation values for the 2-phase case is 0 (the real axis moves from the horizontal to the vertical) while the correlation values for the 4-phase case are similar to those for a phase difference of 0° because of the symmetry of the 4-phase correlations about both axes. This symmetry about both axes indicates that correlation values for the 4-phase Family A set will not be greatly affected by phase differences while the lack of symmetry in the 2-phase Gold set indicates otherwise.

Scatter Plot of Complex Cross-Correlation Values

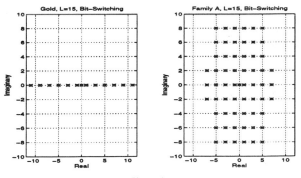

Figure 2

174

V. Probabilistic Performance Analysis

A probabilistic analysis was performed to predict the performance of 2-phase and 4-phase sequences when users have *equal power*. This analysis found the signal-to-noise ratio (SNR) for the two sequence types by dividing the decision statistic's squared mean by its variance. To simplify analysis the following assumptions were made: 1.) random coin toss values were used for the sequences making all chip values Bernoulli(1/2) and independent, 2.) while sequence synchronization was not assumed, chip synchronization (sequence values change at the same time) was used so that mean and variance values could be found per chip and then generalized to sequence sets, and 3.) matched filters were used in that the receiver multiplied the received signal by the basis function of the transmitted signal and integrated over each data bit.

Figure 3 shows the system model used for the 2-phase and 4-phase systems. In both systems the amplitude of the transmitted signal is chosen so that in signal space signals are separated in distance by $2\sqrt{E_c}$ (each user has energy E_c per chip) and the basis function used in the receiver has energy of 1. The derivation of amplitude and other values used in this analysis is done in Appendix II. In the appendix, c refers to the code value, R refers to real, and I to imaginary. Data bit values and bit-switching is not used because sequence values are assumed Bernoulli(1/2) and independent. The first user is assumed to have phase value of 0° to simplify notation; it is the phase differences that are important as seen in later analysis. Also, what is transmitted for the 4-phase system is determined by noting that the receiver only sees the real part of the product of the code, the phase difference, and the modulation.

$$\begin{aligned}
\text{Re}(c_1(t)e^{j\theta_1}e^{jw_ct}) &= \text{Re}[(c_{R1}(t) + c_{I1}(t))(cos\theta_1 + jsin\theta_1)(cosw_ct + jsinw_ct] \\
&= c_{R1}(t)cos(w_ct + \theta_1) - c_{I1}(t)sin(w_ct + \theta_1)
\end{aligned}$$

Using $c_{1I}(t) = 0$ for the 2-phase case gives the correct result.

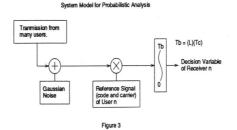

Figure 3

The probabilistic analysis is done in Appendix II. For both 2-phase and 4-phase sequences it was found that the SNR is

$$SNR = (\frac{\mu}{\sigma})^2 = (\frac{N_o}{2E_b} + \frac{S-1}{2L})^{-1},$$

where S is the number of users and is limited to M = L + 2. This result suggests that the performance of 2-phase and 4-phase systems should be equal. Assuming a Gaussian approximation, the P_e for the CDMA system with PSK is found using $P_e = Q(\frac{\mu}{\sigma})$. Note that this analysis assumes chips are independent and Bernoulli(1/2) so the smaller maximum cross-correlation values for 4-phase codes compared to 2-phase codes doesn't come into play. Also of importance is that our probabilistic analysis shows that phase differences don't affect the other-user interference variance (or SNR) for the 4-phase system. In the 2-phase system 90° phase difference produces zero other-user interference variance, while 0° and 180° phase differences cause full other-user interference variance. This result for 2-phase systems will be seen in our deterministic analysis later. For 2-phase systems, Viterbi showed [8] quadriphase modulation by two parallel 2-phase random code sequences causes the other-user interference variance to be independent of phase differences, as in 4-phase systems. The Qualcomm CDMA system incorporates quadriphase modulation by two 2-phase codes [3]. Reducing the amount of interference caused by phase differences is probably desired in CDMA systems with many unsynchronized mobiles.

VI. Exact Performance Analysis

The average probability of error P_e for a 2-user CDMA system with PSK, equal power users, a correlation receiver, and Gaussian noise with variance σ_n^2, can be described as [9]

$$P_e = \frac{1}{N}\frac{1}{C}\sum_r\sum_{q\neq r}\sum_{l=0}^{L-1}\sum_{i=1}^{C}Q[\frac{\mu_{oneuser}}{\sigma_n}(1+I_{r,q,l,i})]$$

where r and q are different code sequences and $I_{r,q,l,i} = \frac{1}{L}[\pm R_{r,q}\pm\hat{R}_{r,q}]$ is the multi-user interference caused by cross-correlations between sequences (which is phase dependent as seen earlier in this paper). $R_{r,q}$ and $\hat{R}_{r,q}$, see Figure 4 which has a 2-phase example, are the partial cross-correlations between the user's code sequence (assuming user's data bit is positive) and the other user's shifted code sequence with bit-switching. The two $\pm$ in $I_{r,q,l,i}$ and the $\sum_{i=1}^{C}$ indicate that data modulation causes different combinations of the partial cross-correlation values. C is the total number of these combinations due to bit-switching. (C = 4 in the two user case because the second user's bits can be both positive(+ +, i=1), both negative(- -, i=2), or one positive and one negative (+ -, i=3 or - +, i=4). Figure 4 shows the (+ -, i=3) combination.) The value l is an integer shift and varies from 0 to $L-1$ based on the simplifying assumption of chip synchronization. $N - (L+2)(L+1)(L)/2$ is the total number of code sequence combinations with integer time shifts. Also of importance is that each data bit is multiplied by an entire code sequence and this code sequence is not changed from data bit to data bit.

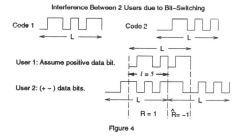

Interference Between 2 Users due to Bit-Switching

Figure 4

A MATLAB program was written to implement the C program in [9]. The plots shown in Figure 5 compare the average probability of error performance for 2 users with 2-phase Gold and 4-phase Family A code sets with sequence lengths of L = 7 and 15. Five curves are shown for each code set because five different runs using random phase values were made to better simulate an actual CDMA system. These random values were different for each choice of sequences r and q but were the same for each code set so that a fair comparison could be made. While the phase values were random, the program still cycles through all code sequences (r and q) and time shifts (l from 0 to L-1).

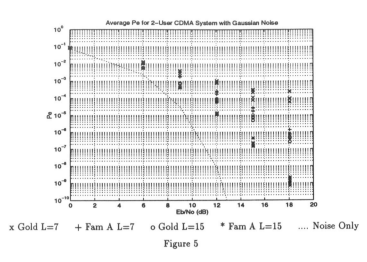

x Gold L=7   + Fam A L=7   o Gold L=15   * Fam A L=15   .... Noise Only

Figure 5

This plot indicates a general trend found in this research, the 4-phase Family A codes offer better average P_e performance than the 2-phase Gold codes, especially at higher values of E_b/N_o where worst-case interference is the limiting factor. This is due to the lower cross-correlation values for the 4-phase code sets compared to the 2-phase code sets, as seen in Figures 1 and 2. The probabilistic analysis done earlier does not take into acccount these lower cross-correlation values achieved with specific code constructions and therefore doesn't predict better error performance for code sets with smaller C_{max} values. It is the C_{max} values that have the greatest impact on average P_e performance because of the sensitivity of the Gaussian $Q(x)$ function to small changes in x. A large negative value of interference $I_{r,q,l,i}$ (from the combination (+ + , - - , + -, or - +) of partial correlations R and $\hat{R}$ due to bit-switching) will cause a large conditional P_e value that will severely effect the average P_e. From this it is important to use code sets with small values of C_{max} with data modulation. Using the same phase differences between all sequences in the code set resulted in plots similar to those shown in Figure 5. For a common phase difference of 90° the 2-user case had better average P_e performance because the multi-user interference was zero. While the 2-phase code set offers better P_e performance for phase differences of 90°, the 4-phase code set offers better phase-averaged P_e performance because of its lower values of C_{max}.

VII. 3-User Performance

To compare the performance of 2-phase Gold and 4-phase Family A code sets for 3 users the MATLAB program for 2 users was altered to include a third user. Due to the long time it would take to cycle through all combinations of code sequences, a MonteCarlo simulation was used. In addition to choosing random phases between users, the three code sequences (r,q,s) were chosen randomly in 7 runs using 50 selections of code sequences and phase differences. These random choices were the same for both code sets for a fair comparison of performance. The output of the program is included in Figure 6 and shows seven different curves for each code set. While the variance of the curves is large due to the limited number of MonteCarlo runs (50) , it is evident that 4-phase Family A sequences offer better average P_e performance than 2-phase Gold sequences. The almost equal performance at low values of E_b/N_o presents a design choice when deciding whether to use 2-phase or 4-phase code sets in an asynchronous CDMA application. The extremely good performance of one of the seven curves for the Gold set illustrates how certain phase differences (such as 90°) can cause other-user interference to be very small for 2-phase systems. While this is true, an average of the results for both code sets (included in the figure) shows that Family A systems still offer better phase-averaged P_e performance than Gold systems especially at higher values of E_b/N_o.

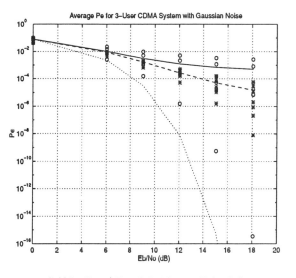

o Gold L=15   * Fam A L=15   .... Noise Only
– Average of Gold Values   - - Average of Family A Values
Figure 6

Appendix I

4-phase set:	2-phase Gold set:
1 0 0 0 3 0 2 3 1 0 3 2 1 3 1	-1 -1 -1 -1 -1 1 1 1 -1 1 1 1 -1 1 1
2 0 0 0 2 0 0 2 2 0 2 0 2 2 2	-1 1 -1 -1 1 1 -1 1 1 1 -1 -1 -1 1 1 1
3 0 0 0 1 0 2 1 3 0 1 2 3 1 3	-1 1 1 -1 1 1 1 -1 1 1 -1 -1 1 1 1 -1 -1 1 1
0 1 0 0 1 3 2 3 2 1 1 1 1 2 2	-1 1 1 1 1 1 1 -1 -1 1 1 1 -1 1 1 1 -1
2 1 0 0 3 3 2 1 0 1 3 1 3 0 0	1 1 1 1 -1 1 1 1 1 1 -1 -1 -1 1 -1 1
3 1 0 0 2 3 0 0 1 1 2 3 0 3 1	-1 -1 1 1 1 -1 -1 1 1 1 -1 -1 1 -1 1 1 -1 1 1 -1
1 2 0 0 1 2 2 1 1 2 1 0 3 3 1	1 1 -1 1 1 -1 -1 -1 1 1 -1 1 1 1 1 1 1
3 2 0 0 3 2 2 3 3 2 3 0 1 1 3	-1 -1 1 1 -1 -1 -1 -1 -1 -1 1 1 -1 -1 1 1 -1 1 1
0 3 0 0 3 1 2 1 2 3 3 3 3 2 2	1 -1 -1 1 1 1 1 -1 -1 -1 -1 -1 -1 -1 1 1 1
1 0 1 0 1 1 1 3 2 2 2 3 0 2 1	-1 1 1 -1 1 1 1 -1 -1 -1 1 1 1 -1 1 -1 -1 -1
3 0 1 0 3 1 1 1 0 2 0 3 2 0 3	1 1 1 -1 -1 1 1 -1 -1 1 1 -1 -1 1 1 1 1 -1
0 1 1 0 3 0 1 3 3 3 0 2 0 1 2	1 -1 1 1 1 1 -1 1 1 -1 1 1 -1 1 1 1 1 -1 1 1
2 2 1 0 2 3 3 0 3 0 3 3 3 1 2	-1 -1 1 1 -1 1 1 -1 1 1 1 -1 1 1 -1 1 1 -1 -1
0 3 1 0 1 2 1 1 3 1 2 0 2 1 2	1 -1 -1 1 1 1 1 1 1 -1 1 1 -1 1 1 1 -1 -1
3 3 1 0 2 2 3 2 2 1 3 2 1 2 1	1 1 -1 -1 -1 -1 1 1 -1 -1 1 1 -1 -1 -1 -1 -1
1 3 2 0 2 3 2 0 1 3 0 3 2 3 3	1 -1 1 1 -1 1 -1 -1 1 1 1 1 1 -1 -1 1 1 1 -1
1 1 3 0 2 2 1 2 2 3 1 2 3 2 3	1 -1 -1 -1 1 1 -1 -1 1 1 1 -1 1 1 -1 1 1 1 1

Appendix II

4-Phase Probabilistic Analysis

The input to the receiver is

$$n(t) + \sqrt{\frac{E_c}{T_c}} \sum_{i=1}^{S} [c_{iR}(t)cos(w_c t + \theta_i(t)) - c_{iI}(t)sin(w_c t + \theta_i(t))]$$

where $n(t)$ is white Gaussian noise, c_i represents the ith user's code sequence, R and I refer to the real and imaginary parts of the 4-phase code, c_{iR} and c_{iI} are Bernoulli(1/2) and independent, and the first user is assumed to have $\theta_1(t) = 0°$ without loss of generality. Using this normalization each user has energy E_c per chip and energy $E_b = LE_c$ per bit. Data bit values and bit-switching is not indicated above because chip values are assumed Bernoulli and independent. The correlation receiver of the first user multiplies this input by the unit energy basis function

$$\sqrt{\frac{1}{T_c}} [c_{1R}(t)cos(w_c t) - c_{1I}(t)sin(w_c t)]$$

and integrates the result from 0 to T_b to get the decision variable Z_i. To find the decision SNR we calculate the mean and the variance of the chip integrals and then find the bit decision SNR using sums of chip variables.

Mean of the decision variable Z_i (μ):

First we find the mean of the chip-duration integral:

$$E(\frac{\sqrt{E_c}}{T_c} \int_0^{T_c} [(c_{1R}(t)cos(w_c t) - c_{1I}(t)sin(w_c t)] \bullet$$

$$[n(t) + \sum_{i=1}^{S} (c_{iR}(t)cos(w_c t + \theta_i(t)) - c_{iI}(t)sin(w_c t + \theta_i(t)))dt])$$

Using the assumption of Bernoulli trials, the uncorrelatedness of the noise and the chip values ($E[n(t)c_{iR}(t)] = 0$), and the independence of sequence values ($E[c_{kR}(t)c_{hR}(t)] = 0$ for $h \neq k$ and $E[c_{iR}(t)c_{jI}(t)] = 0$ for any j) gives $E[.] = \sqrt{E_c}$. The correlation receiver integrates over L chips so $E[Z_i] = \mu = \sqrt{E_c}L$.

Variance due to noise (σ^2_{noise}):

In our system model white Gaussian noise $n(t)$ is added to the CDMA tranmission before multiplication by the basis function in the receiver.

$$\sigma^2_{noise} = Var\left[\sqrt{\frac{1}{T_c}} \int_0^{T_c} [c_{1R}(t)cos(w_c t) - c_{1I}(t)sin(w_c t)] n(t)dt\right]$$

Using the fact that the noise is uncorrelated with the chip values ($E[n(t)c_{iR}(t)] = 0$), the independence of chip values, and properties of white Gaussian noise ($E[n(s)n(t)] = \frac{N_o}{2}\delta(t - s)$), gives

$$Var[.] = \frac{1}{T_c}\frac{N_o}{2} \int_0^{T_c} [cos^2(w_c t) + sin^2(w_c t)]dt = \frac{N_o}{2}.$$

The correlation receiver integrates over L chips so $\sigma^2_{noise} = \frac{N_o}{2}L$.

Variance due to other users (σ^2_{other}):

In addition to the variance from the noise, other-user interference variance is present.

$$\text{Var}\left[\frac{\sqrt{E_c}}{T_c}\int_0^{T_c}[c_{1R}(t)cos(w_ct) - c_{1I}(t)sin(w_ct)]\sum_{i=2}^{S}(c_{iR}(t)cos(w_ct + \theta_i(t)) - c_{iI}(t)sin(w_ct + \theta_i(t)))dt\right]$$

It was found that the mean from other-user interference is 0 so we need to calculate

$$E[\frac{\sqrt{E_c}}{T_c}\int_0^{T_c}[c_{1R}(t)cos(w_ct) - c_{1I}(t)sin(w_ct)]\sum_{i=2}^{S}(c_{iR}(t)cos(w_ct + \theta_i(t)) - c_{iI}(t)sin(w_ct + \theta_i(t)))dt\bullet$$

$$\frac{\sqrt{E_c}}{T_c}\int_0^{T_c}[c_{1R}(s)cos(w_cs) - c_{1I}(s)sin(w_cs)]\sum_{i=2}^{S}(c_{iR}(s)cos(w_cs + \theta_i(s)) - c_{iI}(s)sin(w_cs + \theta_i(s)))ds].$$

In this analysis we assume square pulses so $c_{iI}(t) = c_{iI}(s)$ because s and t are values of time between 0 and T_c. Using independence of chips to get $E[c_{iR}(t)c_{iR}(s)c_{iI}(t)c_{iI}(s)] = 1$ and $E[c_{iR}(t)c_{iR}(s)c_{jR}(t)c_{jI}(s)] = 0$ for $j \neq i$, we find that only certain cross products survive. Retaining only the terms with phase differences (because the double frequency terms integrate to 0) gives:

$$E[X^2] = E(\frac{2E_c}{T_c^2}[\sum_{i=2}^{S}\int_0^{T_c}\frac{1}{2}cos(\theta_i(t) - 0°)dt\int_0^{T_c}\frac{1}{2}cos(\theta_i(s) - 0°)ds]$$

$$+\frac{2E_c}{T_c^2}[\sum_{i=2}^{S}\int_0^{T_c}\frac{1}{2}sin(\theta_i(t) - 0°)dt\int_0^{T_c}\frac{1}{2}sin(\theta_i(s) - 0°)ds])$$

An angle of 0° is subtracted from phase values of all users to indicate that these are phase differences between the first user and all other users. From this it is seen that phase differences between the first user and other users are important, not the specific phases of each carrier. Using the assumption that the phase differences are constant over each chip ($\theta_i(t) = \theta_i(s) = \theta_i$) and $E[sin^2(\theta_i) + cos^2(\theta_i)] = E[1] = 1$ gives

$$E[X^2] = \left[\frac{2E_c}{T_c^2}\frac{1}{4}T_c^2(S - 1)\right].$$

Note that this result shows that the other-user interference variance for 4-phase codes is independent of carrier phase differences between users.

The correlation receiver integrates over L chips so

$$\sigma^2_{other} = L\,E[X^2] = \frac{E_cL(S - 1)}{2}.$$

Combining all of the mean and variance terms for the decision variable and using $E_cL = E_b$ gives the equation for SNR:

$$SNR = (\frac{\mu}{\sigma})^2 = \frac{\mu^2}{\sigma^2_{noise} + \sigma^2_{other}}$$

$$= (\frac{N_o}{2E_b} + \frac{S - 1}{2L})^{-1}.$$

This result indicates that when $\frac{S-1}{2L}$ is much larger than $\frac{N_o}{2E_b}$ system performance is limited more by other-user interference than by channel noise and that the SNR can be modeled as $SNR = \frac{2L}{S-1}$. For $\frac{S-1}{2L}$ much smaller than $\frac{N_o}{2E_b}$ or when no interfering users are present, the SNR can be modeled as $SNR = \frac{2E_b}{N_o}$ as expected.

2-Phase Probabilistic Analysis

The 2-phase probabilistic analysis is very similar to that for 4-phase codes. In the 2-phase case, sequences only have real chip values (± 1) so the I and R subscripts are dropped for easier notation. In both cases white Gaussian noise $n(t)$ is added and $\theta_1 = 0°$ without loss of generality. The input to the receiver is

$$n(t) + \sqrt{\frac{2E_c}{T_c}} \sum_{i=1}^{S} c_i(t) cos(w_c t + \theta_i(t)).$$

Using this normalization each user has energy E_c per chip and energy $E_b = LE_c$ per bit. Data bit values and bit-switching is not indicated above (as in the 4-phase case) because chip values are assumed Bernoulli(1/2) and independent. The correlation receiver of the first user multiplies this input by the unit energy basis function

$$\sqrt{\frac{2}{T_c}}[c_1(t) cos(w_c t)]$$

and integrates the result from 0 to T_b to get the decision variable Z_i. To find the decision SNR we calculate the mean and the variance of the decision variable as before.

Mean of the decision variable Z_i (μ):

First we find the mean of the chip-duration integral:

$$\mathrm{E}\left[\frac{\sqrt{4E_c}}{T_c} \int_0^{T_c} [(c_1(t) cos(w_c t)][n(t) + \sum_{i=1}^{S}(c_i(t) cos(w_c t + \theta_i(t)))dt]\right]$$

Using the assumption of Bernoulli trials, the uncorrelatedness of the noise and the chip values ($\mathrm{E}[n(t)c_i(t)] = 0$), and the independence of chip values ($\mathrm{E}[c_k(t)c_h(t)] = 0$ for $h \neq k$ and $\mathrm{E}[c_i(t)c_i(t)] = 1$) gives $\mathrm{E}[.] = 2\frac{\sqrt{E_c}}{T_c} \int_0^{T_c} cos^2(w_c t)dt = \sqrt{E_c}$. The correlation receiver integrates over L chips so $E[Z_i] = \mu = \sqrt{E_c}L$.

Variance due to noise (σ_{noise}^2):

White Gaussian noise $n(t)$ is added to the CDMA tranmission before multiplication by the basis function in the receiver.

$$\sigma_{noise}^2 = \mathrm{Var}\left[\sqrt{\frac{2}{T_c}} \int_0^{T_c} [c_1(t) cos(w_c t)] [n(t)]dt\right]$$

Using that the noise is uncorrelated with the chip values ($\mathrm{E}[n(t)c_i(t)] = 0$) and properties of white Gaussian noise ($\mathrm{E}[n(s)n(t)] = \frac{N_o}{2}\delta(t - s)$), gives

$$\mathrm{Var}[.] = \frac{2}{T_c}\frac{N_o}{2} \int_0^{T_c} cos^2(w_c t)dt = \frac{N_o}{2}.$$

The correlation receiver integrates over L chips so $\sigma_{noise}^2 = \frac{N_o}{2}L$.

Variance due to other users (σ_{other}^2):

In addition to the variance from the noise, other-user interference variance is present.

$$\mathrm{Var}\left[\frac{2\sqrt{E_c}}{T_c} \int_0^{T_c} c_1(t) cos(w_c t) \sum_{i=2}^{S}(c_i(t) cos(w_c t + \theta_i(t)))dt\right]$$

It was found that the mean from other-user interference is 0 so we need to calculate

$$E[\frac{2\sqrt{E_c}}{T_c}\int_0^{T_c} c_1(t)cos(w_c t)\sum_{i=2}^{S}(c_i(t)cos(w_c t + \theta_i(t)))dt \bullet$$

$$\frac{2\sqrt{E_c}}{T_c}\int_0^{T_c} c_1(s)cos(w_c s)\sum_{i=2}^{S}(c_i(s)cos(w_c s + \theta_i(s)))ds].$$

We assume square pulses (as for 4-phase sequences) so $c_i(t) = c_i(s)$ because s and t are values of time between 0 and T_c. Using independence of chips to get $E[c_i(t)c_i(s)c_i(t)c_i(s)] = 1$ and $E[c_i(t)c_j(s)c_i(t)c_k(s)] = 0$ for $j \neq i \neq k$, we find that only certain cross products survive. Retaining only terms with phase differences (because the double frequency terms integrate to 0) gives

$$E[X^2] = E[\frac{4E_c}{T_c^2}(\sum_{i=2}^{S}\int_0^{T_c}\frac{1}{2}cos(\theta_i(t) - 0°)dt\int_0^{T_c}\frac{1}{2}cos(\theta_i(s) - 0°)ds)]$$

which is different from the 4-phase case (the amplitude is different and there are no $sin[.]$ terms here). An angle of 0° is subtracted from phase values of all users (as in the 4-phase case) to indicate that these are phase differences between the first user and the other users. Using the assumption that phase differences are constant over each chip ($\theta_i(t) = \theta_i(s) = \theta_i$) gives

$$E[X^2] = E[E_c\sum_{i=2}^{S}cos^2\theta_i] = E_c\sum_{i=2}^{S}E[cos^2\theta_i].$$

Note that this result shows that the other-user interference variance for 2 phase codes is dependent on carrier phase differences between users. For the 4-phase case the other-user interference variance was independent of phase differences.

Assuming that $\theta_i \sim U(0, 2\pi)$ we find

$$E[cos^2\theta_i] = \int_0^{2\pi}\frac{1}{2}\frac{1}{2\pi}dt + \int_0^{2\pi}\frac{1}{2}cos(2\theta_i)dt = \frac{1}{2}$$

and

$$E[X^2] = \frac{(S-1)E_c}{2}.$$

The correlation receiver integrates over L chips so

$$\sigma_{other}^2 = L\ E[X^2] = \frac{E_c L(S-1)}{2}.$$

Combining all of the mean and variance terms for the decision variable and using $E_c L = E_b$ gives the equation for SNR:

$$SNR = (\frac{N_o}{2E_b} + \frac{S-1}{2L})^{-1}.$$

This result is the same as for the 4-phase case and shows that a probabilistic analysis, with Bernoulli chip values and averaging over relative carrier phase, indicates that 2-phase and 4-phase codes should have equal performance.

182

References:

[1] S. Boztas, R. Hammons, and P.V. Kumar, "*4-Phase Sequences with Near-Optimum Correlation Properties*," IEEE Trans. Inform. Theory, vol. 38, pp. 1101-1113, May 1992.

[2] J.G. Proakis, *Digital Communications*, 2nd Edition. New York: McGraw-Hill Book Company, 1989.

[3] J.K. Hinderling et al., "*CDMA Mobile Station Modem ASIC*," IEEE Journal of Solid-State Circuits, vol. 28, no. 3, pp. 253-260, March 1993.

[4] L.R. Welch, "*Lower Bounds on the Maximum Cross Correlation of Signals*," IEEE Trans. Inform. Theory, vol. IT-20, pp. 397-399, May 1974.

[5] V. M. Sidelnikov, "*On Mutual Correlation of Sequences*," Soviet Math Doklady, vol. 12, pp. 197-201, 1971.

[6] P.V. Kumar and O. Moreno, "*Prime-Phase Sequences with Periodic Correlation Properties Better Than Binary Sequences*," IEEE Trans. Inform. Theory, vol. 37, pp. 603-616, May 1991.

[7] D.V. Sarwate and M.B. Pursley, "*Crosscorrelation Properties of Pseudorandom and Related Sequences*," Proc. of the IEEE, vol. 68, no. 5, pp. 593-619, May 1980.

[8] A.J. Viterbi, "*Very Low Rate Convolutional Codes for Maximum Theoretical Performance of Spread-Spectrum Multiple-Access Channels*," IEEE Trans. Select. Areas Commun., vol. 8, pp. 641-649, May 1990.

[9] N.L. Ling, "*Quadriphase Sequence For Direct Sequence Spread Spectrum Multiple Access Applications*," Bachelors Thesis at The University of Virginia, August 1992.

Adaptive MLSE equalization forms for wireless communications

Gregory E. Bottomley and Sandeep Chennakeshu

Ericsson, Inc., 1 Triangle Drive, P. O. Box 13969, Research Triangle Park, NC 27709

ABSTRACT

In many mobile communication systems, adaptive equalization is required to combat the effects of multipath time dispersion and a non-stationary channel. In this paper, the principles of maximum likelihood sequence estimation (MLSE) are used to develop an adaptive equalizer, which is an extension of the Ungerboeck receiver. MLSE receivers operating in non-stationary channels require channel estimation and often require channel tracking and prediction. The extended Ungerboeck receiver is shown to require a significant amount of channel prediction. Two alternative MLSE forms are developed which minimize the amount of channel prediction. The first form developed, referred to as the "direct form," uses a standard Euclidean distance metric. A second, novel form, referred to as the "partial Ungerboeck form," provides a lower complexity realization than the "direct form."

1. INTRODUCTION

Digital cellular and PCS systems based on digital AMPS (D-AMPS, IS54) and GSM require an equalizer to handle intersymbol interference (ISI) arising from time dispersion or delay spread in the channel. Typically, a non-linear equalizer such as a decision feedback equalizer (DFE) or maximum likelihood sequence estimator (MLSE) is used in such channels. Changes in the channel from a minimum phase condition to a non minimum phase condition adversely affect the performance of the DFE. Hence, the MLSE approach is usually preferred and has been studied in particular for the D-AMPS channel in [1-4].

The basic approach with maximum likelihood sequence estimation is as follows. Using knowledge of the channel and a hypothesis of the transmitted data sequence, an estimate of the noiseless received sequence is formed, which is then compared to the actual noisy received sequence. The comparison is usually done in an iterative manner by comparing one symbol of the sequence at a time. The hypothesis that provides the best or most likely fit to the received sequence is the detected sequence. The criteria for the "best fit" is dependent on the statistics of the noise process that corrupts the received sequence. Most MLSE receivers used in ISI channels are based on one of two classic approaches provided by Forney [5] and Ungerboeck [6]. In this paper, an adaptive MLSE receiver based on the Ungerboeck approach is developed.

Since the channel is non-stationary, some form of channel tracking using decision feedback is assumed, in which an estimate of the channel is updated using tentative decisions on previously transmitted data symbols . The channel comprises the transmit filtering, which is time-invariant, and a time-varying dispersive propagation medium. For this channel model, the Ungerboeck MLSE form requires a set of channel estimates corresponding to future symbol periods, for each iteration of sequence estimation procedure outlined above. With imperfect channel predictions, as will be the case in a noisy fading channel, the performance will degrade.

To solve this problem, two alternative forms are derived which minimize the need for channel estimation. The first form, referred to as the "direct form" comprises a time-invariant receive filter, matched to the transmit filter, a sampler and a maximum likelihood sequence estimator that operates on the sampled data. The sequence estimator uses a metric which is the Euclidean distance between the filtered received samples and a hypothesis of the filtered received samples. The iterative sequence estimation procedure is realized by the Viterbi algorithm [7]. This form is more complex than the Ungerboeck form, but minimizes the amount of channel prediction. A second more novel form, referred to the "partial Ungerboeck form" is derived from the direct form and also minimizes the amount of channel prediction. This form can be realized with a complexity between the Ungerboeck and direct forms.

This paper is organized as follows. The system model, based on [6], is reviewed in Section II. In section III, the MLSE receiver is derived from the basic principles outlined in [6]. In section IV, the specific case of a time-invariant transmit filter and a time-varying transmission medium is considered, and two alternative forms are derived. Section V concludes this paper.

2. SYSTEM MODEL

The system model follows [6], consisting of a transmitter, transmission medium, and receiver. The receiver converts the radio signal to a complex-valued, baseband signal. The overall channel is assumed to be time-varying. For such a channel, the output $y(t)$ can be related to the input $x(t)$ by means of a time-varying impulse response $h(\tau;t)$. Additive noise $w(t)$ is present at the channel output, so that the received signal is given by:

$$y(t) = \int h(\tau;t) \, x(t - \tau) \, d\tau + w(t) \tag{1}$$

where $w(t)$ is a stationary, white, complex Gaussian noise process in the band of interest, with two-sided power spectral density N_0 and autocorrelation function

$$E\{w(t + \tau)w^*(t)\} = N_0 \, \delta(\tau) \tag{2}$$

assuming down conversion with unity amplitude sinusoids.

The input to the channel is a complex information symbol sequence $\{a_n\}$, so that the input signal $x(t)$ can be expressed as:

$$x(t) = \sum_n a_n \, \delta(t - nT) \tag{3}$$

Substitution of (3) into (1) gives the effective system model:

$$y(t) = \sum_n a_n \, h(t - nT;t) + w(t) \tag{4}$$

While t appears in both arguments of h, it is important to note that the first argument is delay, which gives an impulse response, while the second argument is the time variation of the impulse response. Also, it is assumed that there is a finite time interval over which the received signal is collected, denoted I.

3. UNGERBOECK'S FORMULATION FOR TIME-VARYING CHANNELS

The MLSE receiver finds the hypothetical set of information symbols $\{\alpha_n\}$ that maximizes the likelihood of the received data, given that $\{\alpha_n\}$ were transmitted. This is equivalent to maximizing the log-likelihood function derived from the *a posteriori* distribution of the received signal. Ignoring constant scaling factors and additive terms, the log-likelihood function gives rise to the metric

$$J_H = -\int_{t \in I} \left| y(t) - y_H(t) \right|^2 dt \tag{5}$$

where H is the hypothesis corresponding to the information symbol set $\{\alpha_n\}$ and $y_H(t)$ is the hypothetical received signal [6]. With a time-varying channel,

$$y_H(t) = \sum_n \alpha_n \, h(t - nT;t) \tag{6}$$

First, substituting (6) in (5) and **expanding (step 1)**

$$J_H = -\int_{t \in I} \left(y(t) - \sum_n \alpha_n \, h(t - nT;t) \right)^* \left(y(t) - \sum_k \alpha_k \, h(t - kT;t) \right) dt \tag{7}$$

$$= A + B_H + C_H$$

where

$$A = - \int_{t \in I} |y(t)|^2 dt \tag{8}$$

$$B_H = \int_{t \in I} 2 \operatorname{Re} \left\{ \sum_n \alpha_n^* h^*(t - nT;t)y(t) \right\} dt \tag{9}$$

$$C_H = - \int_{t \in I} \sum_n \alpha_n^* h^*(t - nT;t) \sum_k \alpha_k h(t - kT;t)dt \tag{10}$$

Since term A is independent of the sequence hypothesis H, it can be omitted from the metric.

The next step is to **interchange integration with summation (step 2)** in (9) and (10), so

$$B_H = 2 \operatorname{Re} \left\{ \sum_n \alpha_n^* z(n) \right\} \tag{11}$$

$$C_H = - \sum_n \sum_k \alpha_n^* \alpha_k q(n,k) \tag{12}$$

where

$$z(n) = \int_{t \in I} h^*(t - nT;t)y(t)dt \tag{13}$$

$$q(n,k) = \int_{t \in I} h^*(t - nT;t) h(t - kT;t)dt \tag{14}$$

Substituting $t+nT$ for t, (14) becomes:

$$q(n,k) = \int_{t + nT \in I} h^*(t;t +nT) h(t + (n-k)T;t +nT)dt = s(n - k,n) \tag{15}$$

where

$$s(\ell,n) = \int_{t+nT\in I} h^*(t;t+nT)\, h(t+\ell T;t+nT)\, dt \tag{16}$$

Substituting (15) into (12) gives

$$C_H = -\sum_n \sum_k \alpha_n^* \, \alpha_k \, s(n-k,n) \tag{17}$$

A double summation can be viewed as summing the elements of a matrix. If n is interpreted as a row index and k is interpreted as a column index, then (17) can be interpreted as an inner summation over columns (fixed row) and an outer summation over rows. This is shown graphically in Figure 1a.

An alternative is to split the double sum into a single summation along the matrix diagonal and a double sum, in which partial rows and partial columns are combined. This is shown graphically in Figure 1b. Mathematically, this can be expressed as:

$$\sum_r \sum_c m_{rc} = \sum_r m_{rr} + \sum_r \sum_{\substack{c \\ c < r}} m_{rc} + m_{cr} \tag{18}$$

where the indices r and c refer to row and column indices of the matrix.

The third step is to **re-arrange the double summation (step 3)** in (17), by applying (18) and using the fact that $s^*(k-n,k) = s(n-k,n)$, which gives

$$C_H - -\sum_n \alpha_n^* \, \alpha_n \, s(0,n) - \sum_n \sum_{\substack{k \\ k < n}} 2 \, \mathrm{Re} \left\{ \alpha_n^* \, \alpha_k \, s(n\ k,n) \right\} \tag{19}$$

Finally, the summation over k can be replaced by a summation over $\ell = n-k$, giving

$$C_H = -\sum_n \alpha_n^* \, \alpha_n \, s(0,n) - \sum_n \sum_{\substack{\ell \\ \ell > 0}} 2 \, \mathrm{Re} \left\{ \alpha_n^* \, \alpha_{n-\ell} \, s(\ell,n) \right\} \tag{20}$$

Collecting the results of (20) and (11), the metric can be expressed as

$$J_H = \sum_n M_H(n) \tag{21}$$

where

$$M_H(n) = \text{Re} \left\{ \alpha_n^* \left[2 z(n) - s(0,n) \, \alpha_n - 2 \sum_{\substack{\ell \\ \ell > 0}} \alpha_{n-\ell} \, s(\ell,n) \right] \right\} \quad (22)$$

As discussed in [6], the hypothesis H which maximizes (21) can be determined using a sequence estimation algorithm such as the Viterbi algorithm, in which the one iteration of the algorithm uses the branch metric given by $M_H(n)$, which depends on $\{\alpha_k, k \leq n\}$.

The results in (21) and (22) are similar to the time-invariant result in [6]. The matched filter in [6] becomes the time-varying matched filter in (13). Also, the coefficients in (16) are time-varying forms of the coefficients in [6]. The three steps highlighted in this derivation are basic to deriving the Ungerboeck receiver.

4. WIRELESS COMMUNICATIONS MODEL

Consider a typical wireless model, in which the channel consists of a time-invariant transmit filter, with impulse response $p(\tau)$, followed by a time-varying transmission medium with impulse response $g(\tau;t)$ of the form:

$$g(\tau;t) = \sum_{j=0}^{J-1} g(jT;t) \, \delta(\tau - jT) \quad (23)$$

which is a causal filter with symbol-spaced taps. Such a model gives rise to a symbol-spaced equalizer. The overall time-varying channel response becomes

$$h(\tau;t) = \sum_{j=0}^{J-1} g(jT;t) \, p(\tau - jT) \quad (24)$$

4.1 Ungerboeck form

Substituting (24) into (13) and (16) gives:

$$z(n) = \sum_{j=0}^{J-1} \int_{t \in I} g^*(jT;t) p^*(t - (j+n)T) y(t) dt \quad (25)$$

$$s(\ell,n) = \sum_{j=0}^{J-1} \sum_{k=0}^{J-1} \int_{t+nT \in I} g^*(jT;t+nT) \, g(kT;t+nT) \, p^*(t - jT) \, p(t + (\ell - k)T) dt \quad (26)$$

It is assumed that the time variation of each channel coefficient $g(jT;t)$ is slow relative to the duration of the transmit pulse shape $p(\tau)$. Also, it is assumed that the pulse shape $p(\tau)$ is symmetric and centered at $\tau = 0$. With these assumptions, (25) can be approximated by:

$$z(n) \approx \sum_{j=0}^{J-1} c^*(j;n+j)\, r(n+j) \tag{27}$$

where

$$c(j;n) = g(jT;nT) \tag{28}$$

$$r(n) = \int_{t \in I} p^*(t - nT) y(t)\, dt \tag{29}$$

The same assumptions can be applied to (26), giving

$$s(\ell,n) \approx \sum_{j=0}^{J-1} \sum_{k=0}^{J-1} c^*(j;n+j)\, c(k;n+k-\ell)\, r_{pp}((\ell+j-k)T) \tag{30}$$

where

$$r_{pp}(\tau) = \int p^*(t)\, p(t+\tau)\, dt \tag{31}$$

Assuming the transmit pulse shape convolved with itself gives rise to a Nyquist pulse shape, then

$$r_{pp}(mT) = \delta(m) \tag{32}$$

for integer m. Substituting (32) into (31) gives

$$s(\ell,n) \approx \sum_{j=0}^{J-1} c^*(j;n+j)\, c(j+\ell;n+j) \tag{33}$$

Thus, under these assumptions, the continuous-time received signal $y(t)$ can be first passed through a receive filter matched to the transmit filter, giving rise to a discrete-time signal $r(n)$. Then, an Ungerboeck receiver can be used, operating on the discrete-time signal $r(n)$ with knowledge of the time-varying channel coefficients $c(j;n)$. This is realized using the metric defined by (21), (22), (27) and (33).

At iteration n, the Ungerboeck receiver computes branch metrics $M_H(n)$ for various symbol hypotheses. From (27) and (33), this requires knowledge of the time-varying channel coefficients at times nT through $(n+J-1)T$. This knowledge may be difficult to obtain in practice if decision feedback is being used to estimate the channel, as discussed in [8]. Typically, after iteration n is completed, tentative decisions on symbols $s(n-d)$ are made, where d is an update decision delay. Using these tentative decisions and channel knowledge at time $n-d$, an expected $r(n-d)$ value can be generated and subtracted from the received $r(n-d)$ value, generating an error signal. This error signal can be used to update the channel coefficient estimates, giving values corresponding to time $n-d+1$. The channel coefficients must then be predicted or extrapolated to time $n+1$, the next iteration of the equalizer [9]. With the Ungerboeck form, predictions would be needed at times $n+1$ through $n+J$, requiring d-step through $(d+J-1)$-step prediction. The baseband processor of the receiver is illustrated in Figure 2.

With typical channel estimation and prediction algorithms, the accuracy of the prediction decreases with the number of steps over which the prediction is made. Thus, from a practical implementation point of view, there is interest in a receiver form which minimizes the prediction of channel coefficients.

4.2 Direct form

One form which minimizes prediction, referred to as the "direct form," can be derived from the Ungerboeck form. Substituting (27) and (33) into (11) and (17), respectively, gives

$$B_H = 2\,\mathrm{Re}\,\left\{ \sum_n r_H^*(n)\, r(n) \right\} \tag{34}$$

$$C_H = -\sum_n \left| r_H(n) \right|^2 \tag{35}$$

where

$$r_H(n) = \sum_{j=0}^{J-1} c(j;n)\, \alpha_{n-j} \tag{36}$$

Completing the square, i.e., adding a summation over n of $|r(n)|^2$, which does not impact which hypothesis is selected, the resulting metric is given by (21) where

$$M_H(n) = -\left| r(n) - r_H(n) \right|^2 \tag{37}$$

The direct form processes the filtered received data samples $r(n)$ sequentially. At iteration n, branch metric $M_H(n)$ depends only on channel coefficients at time n, as seen in (36) and (37), minimizing the need to predict channel coefficients. The direct form is well known in the wireless industry, and the above derivation provides only a formal development of the receiver. While this form minimizes the need for prediction, it requires more computations per iteration than the Ungerboeck form. The direct form is illustrated in Figure 3.

4.3 Partial Ungerboeck form

A second alternative form is obtained from the direct form by applying two of the three steps used to derive the Ungerboeck form. The first step is performed by expanding the metric in (21) and (37) and dropping the constant term, which leaves the terms in (34) and (35). Also, (36) is substituted into (34) and (35), giving

$$B_H = \sum_n 2\,\mathrm{Re}\left\{ \sum_{j=0}^{J-1} \alpha_{n-j}^* Z(n,j) \right\} \tag{38}$$

$$C_H = -\sum_n \sum_{j=0}^{J-1} \sum_{k=0}^{J-1} c^*(j;n)\, c(k;n)\, \alpha_{n-j}^* \alpha_{n-k} \tag{39}$$

where

$$Z(n,j) = c^*(j;n)\, r(n) \tag{40}$$

The second step, which would be interchanging two summations in this case, is not performed, as this would move the summation over n inside the j and k summations in (39), effectively re-ordering terms so that future channel coefficients would be needed.

A modified form of the third step is applied, in which the expression in (18) is used with $c<r$ being replaced by $c>r$. As a result, (39) becomes:

$$C_H = -\sum_n \sum_{j=0}^{J-1} \alpha_{n-j}^* \left(S(0,n,j)\alpha_{n-j} + 2 \sum_{\ell=1}^{J-1-j} S(\ell,n,j)\, \alpha_{n-j-\ell} \right) \tag{41}$$

where

$$S(\ell,n,j) = c^*(j;n)\, c(j+\ell;n) \tag{42}$$

Putting these results together, the accumulated metric is given by (21), in which

$$M(n) = \sum_{j=0}^{J-1} \mathrm{Re}\left\{ \alpha_{n-j}^* \left[2\,Z(n,j) - S(0,n,j)\,\alpha_{n-j} - 2\sum_{\ell=1}^{J-1-j} \alpha_{n-j-\ell}\,S(\ell,n,j) \right] \right\} \tag{43}$$

This form is similar to the Ungerboeck form, except that the summation over j appears on the outside of the branch metric, rather than the inside. Because only some of the steps used in deriving the Ungerboeck receiver are employed, this form is referred to as the "partial Ungerboeck form." The partial Ungerboeck form, illustrated in Figure 4, has a complexity which is less than the direct form, but greater than the Ungerboeck form.

4.4 Further simplification

For two of the forms, further simplification is possible if the modulation is a form of PSK, i.e., the magnitude of all possible digital symbols is the same. For the Ungerboeck form, the $s(0,n)\alpha_n$ term may be omitted as well as the factors of 2 in (22). For the partial Ungerboeck form, the $S(0,n,j)\alpha_{n-j}$ term may be omitted as well as the factors of 2 in (43).

5. CONCLUSION

The Ungerboeck receiver derivation was extended to the case of a time-varying channel. A typical wireless channel model was considered, in which the channel consists of a time-invariant pulse shaping filter at the transmitter followed by a time-varying transmission medium. It was shown that the Ungerboeck form may have a disadvantage when practical channel estimation is used, as a significant amount of channel prediction is required. Two alternative receiver forms were developed which minimize the need for channel prediction, but which require more complexity.

REFERENCES

[1] G. Larsson, B. Gudmundson, and K. Raith, "Receiver performance of the North American digital cellular system," *Proceedings VTC '91*, St. Louis, MI, May 19-22, 1991.

[2] R. D. Koilpillai, S. Chennakeshu and R. T. Toy, "Low complexity equalizers for U.S. digital cellular system," *Proceedings VTC '92,* Denver, CO, May 10-13, 1992.

[3] Q. Liu and Y. Wan, "An adaptive maximum-likelihood sequence estimation receiver with dual diversity combining/selection," *Intl. Symp. on Personal, Indoor and Mobile Radio Commun.*, Boston, MA, pp. 245-249, Oct. 19-21, 1992.

[4] P. K. Shukla and L. F. Turner, "Examination of an adaptive DFE and MLSE/near-MLSE for fading multipath radio channels," *IEE Proc.-I,* vol. 129, no. 4, pp. 418-428, Aug. 1992.

[5] G. D. Forney, "Maximum-likelihood sequence estimation of digital sequences in the presence of intersymbol interference," *IEEE Trans. Info. Theory,* vol. IT-18, pp. 363-378, May 1972.

[6] G. Ungerboeck, "Adaptive maximum likelihood receiver for carrier modulated data transmission systems," *IEEE Trans. Commun.*, vol. COM-22, pp. 624-635, May 1974.

[7] G. D. Forney, Jr., "The Viterbi algorithm," *Proc. IEEE,* vol. 61, pp. 268-278, Mar. 1973.

[8] J. G. Proakis, *Digital Communications, 2nd ed..* New York: McGraw-Hill, 1989.

[9] E. Dahlman, "New adaptive Viterbi detector for fast-fading mobile-radio channels," *Electr. Lett.*, vol. 26, pp. 1572-1573, Sept. 1990.

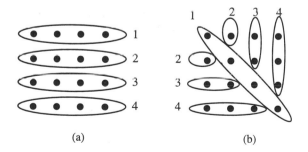

(a)   (b)

Figure 1. Methods for summing elements in a matrix.

194

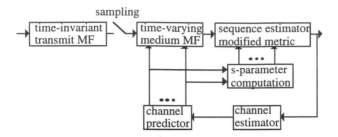

Figure 2. Ungerboeck receiver for special case.

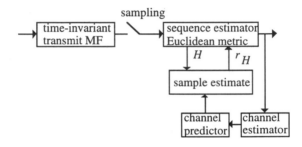

Figure 3. Direct form receiver for special case.

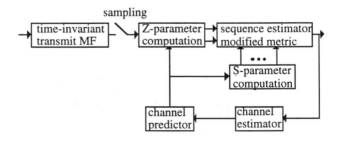

Figure 4. Partial Ungerboeck receiver for special case.

Simulation of RBDS AM Subcarrier Modulation
Techniques to Determine BER and Audio Quality

Shaheen Saroor

University of Wyoming, Electrical Engineering Dept.

Robert Kubichek

University of Wyoming, Electrical Engineering Dept.
Laramie, WY 82071-3259 USA

Jim Schroeder

University of Denver, Dept. of Engineering, Knudson Hall
2390 S. York, Denver, CO 80208-0177 USA

Jim Lansford

Mobile Data Systems, P. O. Box 60488
Colorado Springs, CO 80960 USA

Abstract - This paper describes the modeling and simulation of candidate AM subcarrier systems designed to transmit digital information for display on the dial of next generation radios. AM subcarrier technology still remains under-developed and its possible applications untapped. One such application of AM subcarriers is the Advanced Traffic/Traveler Information System. In the proposed system, the digital signal is inserted between the AM carrier and the audio sidebands. This paper investigates the effect of mutual interference between the audio and digital programs for a variety of digital modulation schemes, bit rates, and channel conditions. The GMSK subcarrier model exhibited the best overall performance in terms of spectrum efficiency and output signal-to-noise ratio.

I. INTRODUCTION

With FCC deregulation of subcarrier usage, broadcasters and communication industry have moved ahead in researching and developing radio broadcast data system (RBDS), that enable radio stations to transmit data for display on the dial of next generation radios. RBDS allows stations to provide secondary services via subcarrier including stock quotations, paging services, advertising messages, and road and travel information (for example as part of the Advanced Traffic/Traveler Information System (ATIS)).

The bulk of recent research and development in subcarrier technology has been limited to use in FM broadcasting systems. Even though FM stations far outnumber AM stations, they are conspicuously scarce in many rural areas. In states such as Wyoming, Idaho, Montana, and Colorado FM coverage is very limited - especially in mountainous or remote areas. Only AM subcarriers can provide consistent wide coverage needed for effective RBDS applications such as the rural ATIS system.

Administrators and researchers alike have taken the position that Intelligent Transportation Systems (ITS) is to be supported through an architecture of hybrid physical - layer technologies, including RBDS subcarrier transmission. Accordingly, at a recent ITS meeting held in April 1994 in Atlanta, the Federal Highway Administration, the Dutch Ministry of Transport, and the Ministry of Transportation of Ontario, announced the award of Herald, an en-route driver advisory system that will use AM subcarrier for information transmission.

One such AM subcarrier scheme was pioneered in 1982 by Institut für Rundfunktechnik (IRT), German Post Office, and BBC [9]. It broadcasted digital data at 100 bits per second. using the BPSK modulation of the AM carrier with the phase shift limited to ±15 degrees. The transmitted digital data was intended for traffic news and navigation aid for drivers. The receiver associated with this subcarrier system required additional processing circuit, a speech synthesizer, a temporary store for traffic announcements, a fixed store for navigation, a display, and also an optional printer.

In this paper we examine a number of alternatives to the ±15 degrees BPSK system with the goal of increasing the bit rate to 200 bits per second or more, while maintaining the quality of the audio program. In the proposed system, the digital signal is inserted between the AM carrier and the audio sidebands. The effect of mutual interference between the audio and digital signals for a variety of digital modulation schemes, bit rates, and channel conditions are examined. Performance benchmarks such as high output signal-to-noise ratio (SNR) of the received analog speech signal and low bit error rate (BER) of the received RBDS signal, governed our inference of best performance, effectiveness and the feasibility of candidate AM subcarrier systems.

The ACOLADE software package was used exclusively to develop and simulate the subcarrier models, while MATLAB was used to analyze ACOLADE simulation data. All simulations were carried out at baseband.

II. AM Subcarrier Model

In order to provide subcarrier services, secondary signals like RBDS, are located above or below the audio signal spectrum. In FM transmission for example, the data subcarrier is located between 59 and 75 Khz or 6 Khz above the highest stereo frequency. In AM systems, the broadcast analog signal is typically high-pass filtered with a cut off of 50 - 100 Hz. This leaves about 100 - 200 Hz of bandwidth in the passband AM signal, that can be used to transmit the digital signal. The idea is to pack the modulation spectrum between the carrier and the audio sidebands, and keep interference between digital and analog signal to a minimum. Fig. 1 shows a conceptual model of the passband spectrum of the composite signal. The basic AM subcarrier model developed and simulated using ACOLADE is shown in Fig. 2.

Five digital modulation schemes were simulated including BPSK, QPSK, GMSK, MSK, and QAM. The audio program was simulated using male speech sampled at 8000 samples per second and high pass filtered with 100 Hz cutoff frequency. The channel was selected as a simple added - gaussian noise model where channel SNR was varied in a specified range. Digital BER for each SNR was computed through ACOLADE simulation. The audio was demodulated using envelope detection and saved for post analysis.

It is critical to the success of AM subcarrier applications that audio quality not be compromised by the presence of the digital subcarrier. Such interference effects were studied by both informal listener tests, and by measuring the output audio SNR. Prior to estimating the output SNR, the received speech signal was processed through a 100 Hz high pass filter to simulate the response of typical AM receivers. Output SNR was estimated using a modified version of the algorithm given by [4],

$$SNR = E^2(S^2) / (E(X^2)E(S^2) - E^2(SX)).$$ (1)

where $E(*)$ is the expectation operator, and S and X are the input and output signals respectively.

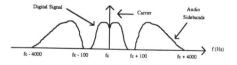

Fig. 1. Spectrum of the composite AM subcarrier signal.

III. Results

The output SNR of the received audio and the BER of the received digital signal were computed for each candidate digital modulation scheme. In analog signals, the chief objective is the fidelity of reproduction of waveforms, making output SNR a suitable criteria of system performance. Output SNR versus bit rate is shown in Fig. 8. This plot indicates that GMSK modulation produced the highest quality audio by a large margin, especially for bit rates ranging from 100 to 200 bits per second. Informal listening tests verify these results in that distortion is detectable as a low 'hum' only at bit rates above 300 bits per second. All other modulation schemes produced audible distortion at all tested bit rates. The presence of the digital signal can affect the audio component in two ways. First, spectral components leak onto the analog sidebands where they are heard as noise. Secondly, and perhaps more seriously, when the digital signal is vector summed with DSB+C audio component, the result can appear as an overmodulated signal at the input to the envelope detector. As an example the input and distorted audio spectrum for QPSK modulation at 100 bits per second is shown in Fig. 9. Significant distortion is observed at all frequencies. Compare this to the GMSK case at 100 bits per second in Fig. 10, which shows no distortion. The effect of GMSK can only be seen in the 0 - 100 Hz range.

Bit errors in digital program are caused by added noise in the channel as well as energy from the audio sidebands that leaks onto the digital spectrum. Increased interference is expected for modulation schemes with poor spectral efficiency. Higher bit rates will also result in more overlap of the audio and digital spectra, further increasing BER values. Figures 3 - 7 show BER versus channel SNR for a range of bit rates. It is noted that simulated SNR values are lower than would be expected in typical operating conditions (usually > 20 dB). This was necessitated by long simulation time required to estimate BER at high SNR. In general the curves show that the audio program caused increased BER, especially at higher bit rates. The exception is GMSK, where BER values are shown to be lower than the ideal (no audio present) case. Presumably this is due to statistical variations in the simulation results. The GMSK BER values were higher than for the other schemes, resulting in error rate of 10^{-1} and above for bit rates over 200 bps. Better error rates can be expected for channel noise of 20 to 30 dB SNR which is more representative of typical AM operation.

IV. Conclusion

This paper examined the practicality of an AM subcarrier system. Five different digital modulation schemes were simulated over a range of bit rates and channel noise conditions. The GMSK system proved superior to BPSK, QAM, QPSK, and MSK in terms of BER and audio output SNR for data rates up to 200 bits per second. Additional informal listening test indicated that the GMSK subcarrier signal produced no detectable interference in the audio program. These preliminary results indicate that an AM subcarrier is indeed a viable solution for providing 200 bits per second digital information services such as rural ATIS.

Current research is focused in the following areas. First, BER performance curves being computed for higher and more practical SNR values. Additional digital modulation schemes are being investigated, for example pi/4 DQPSK with pulse shaping may provide even better performance. Finally, future channel models will include effects of fading and multipath.

REFERENCES

[1] B. P. Lathi, *Modern digital and communication systems*, 2nd ed., Holt, Rinehart and Winston, Inc. 1989.

[2] I. Korn, *Digital communications*, Van Nostrand Reinhold Company. New York: 1985.

[3] J. P. M. G. Linnartz, "Spectrum efficiency of radio data systems (RDS)," *IEEE Trans. On Broadcasting*, vol. 39, pp. 331-334, September 1993.

[4] M. C. Jeruchim, P. Balaban, and K. S. Shanmugan, *Simulation of communication systems*, Plenum Press. New York: 1994.

[5] S. Goldman, and J. Crumbacher, "Two dozen ways you can profit from Subcarriers," *Personal Comm. Magn.* pp. 10-12, July-August 1985.

[6] S. Pasupathy, "Minimum shift keying: A Spectrally efficient modulation," *IEEE Comm. Magn.* vol. 17, pp. 14-22.

[7] P. Scomazzon, "Comparative study of digital modulations accompanying VHF/FM broadcasts," *EBU Review, Technical*, vol. 249, pp. 170-183, August 1991.

[8] P. L. Mothersole, and N. W. White, *Broadcast data systems*, Butterworth & Co. Ltd. London: 1990.

[9] R. Deuscher, "Medium and long waves: Data besides sound broadcasting," *Funkshau*, vol. 9, pp. 48-51, April 1990.

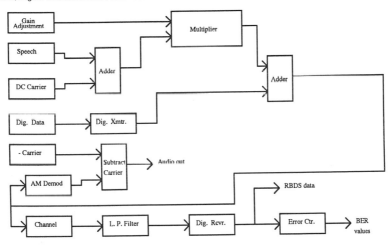

Fig. 2. ACOLADE AM subcarrier model. Note - simulation is carried out using complex baseband signals. Therefore no oscillator or mixer component are required. Also note rethat the Gaussian channel was applied to the digital data for estimating BER performance, but was not used for the audio system.

198

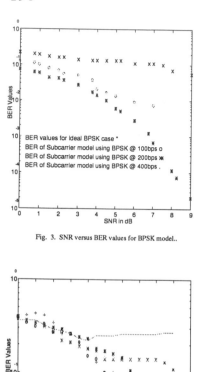

Fig. 3. SNR versus BER values for BPSK model..

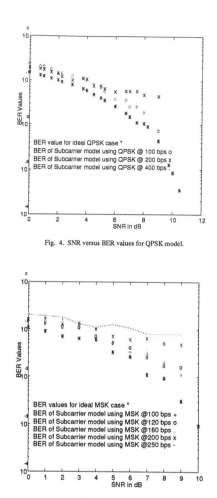

Fig. 4. SNR versus BER values for QPSK model.

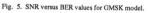

Fig. 5. SNR versus BER values for GMSK model.

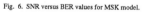

Fig. 6. SNR versus BER values for MSK model.

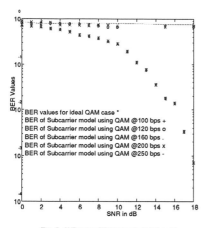

Fig. 7. SNR versus BER values for QAM model.

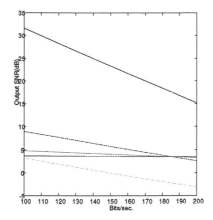

Fig. 8. Output SNR versus bit rate plot for various models. From top to down: Plots of GMSK, QAM, QPSK, BPSK, and MSK models respectively.

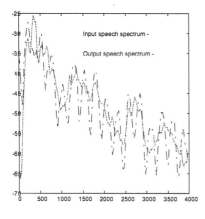

Fig. 9. Input and output speech spectrum of QPSK model at 100 bits per second (bps).

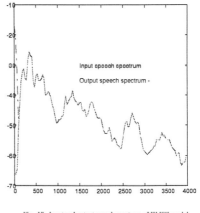

Fig. 10. Input and output speech spectrum of GMSK model at 100 bits per second (bps).

200 Blank

Simulation of Reverse Channel of Narrowband PCS

Rade Petrovic, John N. Daigle, Paolo Giacomazzi[1]

Center for Wireless Communications, University of Mississippi

Abstract

The reverse channel of narrowband PCS is shared among all of the portable units, which use the channel to access a collection of base receivers. Multiple portable units may transmit simultaneously in the same frequency channel, giving rise to strong co-channel interference. The movement of portables causes variations in signal powers, as well as phase noise. The combined effects of noise, fading, co-channel interference, shadowing and other phenomena have not been exhaustively considered in the literature. In this paper, we discuss development of models of multiple co-channel interference based on measurements of single and double co-channel interferers. We then describe a Monte Carlo simulation through which one can assess packet success rate and other performance measures as a function of important system parameters, such as the modulation scheme, the signaling rate, and the error-correction code deployed. We also provide a collection of simulation results and discuss their implications.

1 Introduction

In August 1994 , the FCC auctioned a number of 50 KHz channels for the deployment of nationwide Narrowband Personal Communications Systems (NPCS). Services developed under these licenses are intended to provide two-way communications to serve a general population of mobile subscribers. In a typical implementation, users would communicate with people on the fixed part of the network and with each other via a network of base transmitters (BTs) and base receivers (BRs). The BTs would transmit messages in the forward direction, from the network to the personal messaging units (PMUs) carried by the subscribers, and the BRs would receive messages from the PMUs to the network.

Transmissions of all BTs in a given geographic area are simulcast, but the problem of sharing the reverse channel efficiently would typically involve simultaneous transmission of packets from different users at the same time over the same frequency channel and in the same geographic area, giving rise to strong co-channel interference (CCI). In many systems, subscribers share reverse channel resources using a version of the slotted-Aloha protocol [1]; one example is Motorola's Altair indoor wireless communication networking product line [2], which uses a variation on slotted-Aloha with reservation. In both the Motorola system just mentioned and in the NPCS, part of the reverse channel is organized as a sequence of contention slots, and the users contend for reverse channel resources by transmitting reservation packets in those slots. An alternative approach is to organize the entire reverse channel as a sequence of contention slots and have the subscribers use the reverse channel by transmitting their messages as a sequence of fixed-length packets.

One determinant of the quality of service perceived by the users is the number of successful packets per slot ($SPPS$), the average number of distinct packets received successfully by the network of BRs, as a function of the number of transmissions that occur within a geographic area during a given time slot. The objective of this paper is to present the results of an analysis of the $SPPS$ for a predefined set of system parameters including modulation technique, packet length, forward-error-correction (FEC) code, transmitter power, receiver sensitivity, and antenna heights.

The crucial step in estimating $SPPS$ is to develop a model of CCI that adequately reflects system behavior. A comprehensive model should take into account the number of simultaneous

---

[1]On leave from the Politecnico di Milano, Italy. Authors' names are listed in the order in which the investigators joined the project.

transmissions and the combined effects of noise, fading, shadowing, and position and speed of movement of the PMUs. There is a wealth of literature on this subject, for example [3-9], but our experience suggests that there is still room for improvement.

In Section 2, we provide a brief description of the specific NPCS upon which our analysis is based. In Section 3, we present laboratory measurement results of the packet success rate (PSR), the probability that the target packet transmission from a particular PMU is received successfully at a specific BR, which we will denote by $PSR(PMU, BR)$, in the presence of Rayleigh fading both with and without CCI. Our measurements indicate that in the presence of Rayleigh fading at speeds of movement between 3 and 100 kilometers per hour there is actually a transition region of more than 13 dB between the signal-to-noise ratio (SNRs) below which PSR is less than 0.1 and above which PSR is greater than 0.9. Indeed, there is a transition region of more than 4 dB even in the case in which the received power remains constant over the period of a packet transmission. This empirical result differs from what is customarily assumed in the literature, where the general practice is to define a threshold, R, such that if the ratio of received-target-signal power to noise power exceeds R, then successful reception of the target signal is assured.[2] Indeed, our results suggest that the transition region for PSR is even larger in the case of CCI, where the threshold effect is also typically applied in the literature.

A little thought will reveal that it is significantly more costly to obtain meaningful measurements of PSR in the case of multiple independent interferers than in the case of a single interferer; for example, multiple channel simulators, multiple PMUs, and (more importantly) multidimensional measurements would be needed. We address the issue of obtaining suitable expressions of PSR for the case of multiple CCI and additive white Gaussian noise (AWGN) in Section 4.

In the literature [3-9], the threshold, or *capture*, model is typically applied across the board and the effects of multiple interferers are accounted for in one or more of the following three ways: (1) substitute a single interferer whose power is equal to the sum of the powers of the individual interferers[3, 8, 9] (we call this the *sum* model), (2) substitute a single interferer with power equal to the maximum power of the individual interferers (we call this the *strongest interferer only* model), or (3) treat successful transmission in the presence of I interferers as the intersection of I independent events each of which has a probability of success determined solely by the effect of a single interferer (we call this the *product* model). It is claimed in [3], without proof, that alternative (2) provides a lower bound on capture probability; more will be said on this later.

A typical approach to treating AWGN is to add the noise power to the total interference power and then apply the capture model as in (1) of the previous paragraph. Our measurements show that this approach also yields optimistic results; the AWGN actually produces more packet errors than interference of the same power. This is especially true in a fading environment. Again we will propose a model that combines AWGN and CCI and fits our measurements.

In Section 5, we present a simulation model through which we calculate the number of successful packets per slot ($SPPS$) for a given base receiver network, which is a function of the PSR for each transmitting PMU at each BR in the network of BRs. Averaging is done over repeated events with randomly chosen PMU positions and speeds of movement, shadowing, and building penetration. A byproduct of the analysis is the *hit rate*, the average number of successful receptions per packet transmission for a given network of BRs. The simulation results, which are used to predict the performance of a given BR network configuration, are discussed in Section 6. We believe these results will be an important factor in optimizing network design.

Finally, conclusions and a limited discussion of open issues are presented in Section 7.

---

[2]In the literature, the threshold, R, may be applied to either a message or a bit, the former being referred to as *slow* fading and the latter as *fast* fading.

2 Description of the NPCS System Under Consideration

The NPCS considered here is a narrowband, two-way messaging system. The forward channel, from the base transmitters to the PMUs, operates in a simulcast mode at about 25 Kbps. In the reverse channel, the PMUs transmit at about 10 KBps using 4-FSK. Pre-modulation pulse shaping for the reverse channel is accomplished using a second order Butterworth filter, and the carrier is in the 940-MHz band.

Transmission in the reverse channel is synchronized to the forward channel and organized as a sequence of time slots, each occupying about 16 ms. Each repeating cycle of the reverse channel contains a group of slots used for scheduled transmission, which is controlled by the forward channel, followed by a group of slots allocated to contention among the subscribers, who contend according to a variation of the slotted-Aloha protocol. The Aloha portion provides many functions, among which are PMU registration, transmission reservation, and certain types of acknowledgments. During the period of scheduled transmissions, a network controller can allocate slots to PMUs based on user location information in order to minimize CCI, and multiple-slot allocations are possible.

In this paper we consider only the Aloha portion of the reverse channel, and all packet transmissions have a single-slot duration. The single-slot packets are contained in the payload of a Reed-Solomon codeword, which includes 40 bits of error-correction overhead. Since the error-correction code, which is a (31,23) Reed-Solomon code, can correct up to four symbol errors of five bits each, the code will always correct packets having four or fewer bit errors and can correct up to 20 bit errors.

3 Measurement System and Results

We now discuss our laboratory test set up. First, a target packet is created using a simulated PMU. This packet is down-loaded into a function generator which generates the four-level signal by playing out the recorded packet repeatedly. Pulse shaping is accomplished by an analog filter, whose output is used to modulate a signal generator. Fading is simulated by an HP-11759 RF-channel simulator. The signal is then passed through a hybrid to facilitate introduction of interference, then through an attenuator to control received signal power. Following demodulation and detection, a bit error rate analysis is performed on the codeword representing the packet. Next, RS decoding takes place and finally packet error rate analysis is accomplished. In our tests we used three different versions of BRs, one was designed at the University of Mississippi and the other two were beta test products from Glenayre and Motorola. All of the receivers are of the noncoherent, limiter-discriminator type, and they all showed very similar sensitivity performance.

Typical results of PSR vs. signal power in the Rayleigh fading environment for different speeds of movement are given in Fig. 1. As a reference we plotted the results for a static (no fading) environment. From Fig. 1 we see that a transition from $PSR = 0.1$ to PSR $= 0.9$ typically occurs over a signal-strength increase of about 14 dB. It should be noted that an increase in speed of movement reduces the receiver sensitivity significantly. This can be explained by an increase in FM-click density and amplitude [10]. This figure also indicates there is a transition region of more than 4 dB even in the case in which the received power remains constant over the period of a packet transmission (static case), although the common assumption is that this transition takes place over a range that can be assumed to be approximately 0 dB.

In order to measure the effects of CCI, we generated an interference packet and used a second function generator to produce an interfering signal synchronized to the desired signal. The RF channel simulator was used to fade the target and interference signals independently. Both signals were strong so that the effect of noise can be considered negligible. Typically the desired signal was -50 dBm while the interference was varied in order to produce different signal-to-interference ratios (SIRs).

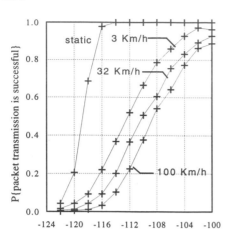

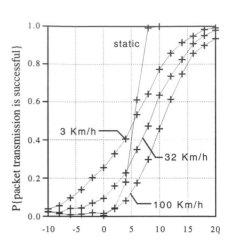

Signal Power [dBm]   Signal-to-Interference Ratio [dB]

Figure 1: Probability that packet transmis-   Figure 2: Probability that packet transmission
sion is successful as a function of SNR under   is successful as a function of SIR under Rayleigh
Rayleigh fading at different speeds of mobility. fading at different speeds of mobility.

Fig. 2 shows the PSR vs. SIR for different fading conditions. Without fading the transition from a PSR of 0.1 to a PSR of 0.9 occurs over an SIR range of only 4 dB. In the presence of Rayleigh fading the transition takes place over a 20 dB range at low speeds of movement, and the transition region decreases as speed of movement is increased. As noted earlier, the higher the speed, the lower PSR is measured.

As Fig. 2 shows, it is possible to receive the packet correctly even for a negative SIR. Recall that SIR is defined as the ratio of the long-term average received target signal power to the long-term average interference power. But, at any point in time, it is possible that the received signal power may exceed the received interference power even if the SIR is negative. At low speeds, it is possible that this favorable condition may prevail over the entire period of the target-packet transmission. At higher speeds, it is still likely that the received signal power to received interference power may be favorable at some time during the transmission, but it is less likely that the favorable conditions will last throughout the period of target-packet transmission. Therefore, successful transmission is less likely at higher speeds and the transition region is steeper at high speed than for low speeds.

It should also be noted that if the interference signal is itself a packet transmission, then the effective PSR at a given node is obtained by summing the PSR for each of the competing trans-missions. That is, if the SIR as seen by one competing transmission is say, -6 dB, then the SIR as seen by the remaining competing transmission is +6 dB. As an example, at 3 Km/h the PSR at an SIR of -6 dB would be approximately 0.1 + 0.55 so that in the case of two competing packet trans-missions, the overall success rate would be 0.65. For this reason, we distinguish $PSR(PMU, BR)$ and $PSR(BR)$.

Comparison between Figs. 1 and 2 shows that the effects of fading are very different for interference and noise. First, fading generally reduces the PSR in presence of noise, while in the interference-only case the fading gives an improvement for small SIRs, especially for low speeds of movement. Second, the transition region is steeper in noise than in CCI, and the effects of speed are different. This can be explained by the fact that the interfering signal is subjected to fading while the noise is not. Interfering signal and noise amplitudes are also drawn from different distributions: Rayleigh and Gaussian, respectively. Therefore, the usual practice of adding the powers of interference and noise in a simulation may be justified for static case, in which the noise and interference powers do not change over the period of packet transmission, but not in a fading environment, where a more representative model is needed.

Fig. 3 shows results for PSR in the presence of two independently faded signals with different SIRs: SIR1 and SIR2. SIR1 is given on the abscissa while SIR2 serves as a parameter for the different curves. First, it can be observed that SIR2 must be increased to about 30 dB in order to obviate the effects of the second interferer. For the purposes of further analysis, the curve $PSR(SIR1)$ for $SIR2 = 30$ dB can be considered as representative of the single-interferer case.

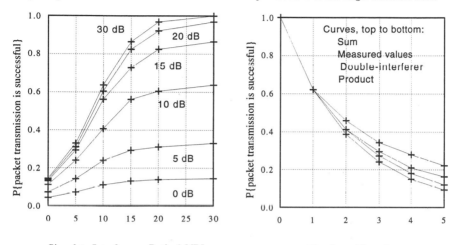

Signal-to-Interference Ratio 1 [dB]

Number of Interferers

Figure 3: Probability that packet transmission is successful in the presence of two interferers as a function of SIR1 with SIR2 as a parameter.

Figure 4. Probability that packet transmission is successful in the presence of multiple interferers as a function of the number of interferers.

A closer observation of results given in Fig. 3 leads to the conclusion that the PSR for all measured points can be calculated approximately by the formula

$$PSR(SIR1, SIR2) = PSR(SIR1)PSR(SIR2), \quad (1)$$

where $PSR(SIR1)$ and $PSR(SIR2)$ are both taken from the 30-dB $PSR(SIR1)$ curve. We will call the model represented by formula (1) the *product* model. A detailed examination of the product model shows that the biggest discrepancies between formula (1) and our measured results occur for small values of $SIR1$ and $SIR2$, where formula (1) underestimates PSR. For example for $SIR1 = SIR2 = 0$ dB, we measured $PSR(0, 0) = 0.043$. However, we measured $PSR(0, 30) = 0.144$ while formula (1) gives $PSR(0, 0) = 0.021$, a pessimistic prediction.

An improved match to measured results is obtained by the following formula:

$$PSR(SIR1, SIR2) = PSR(SIR1)PSR(SIR2)^{PSR(SIR1)^{0.24}} \quad (2)$$

where $SIR1 > SIR2$.

4 Multiple Interference Models

We mentioned earlier that a typical approach to determining PSR in the literature is to hypothesize the existence of a ratio of received target signal power to interferer power above which successful transmission of the target packet is assured. Mathematically,

$$PSR = P\{X/Y > R\} \qquad (3)$$

where X and Y represent the target-signal and interference powers, respectively, and R represents the threshold. Upon conditioning on the value of the interference power and assuming that the target signal value is drawn from the exponential distribution with parameter α as in slow Rayleigh fading, we find

$$PSR = \int_0^\infty P\{X > Ry\}\, dF_Y(y) = \int_0^\infty \exp\{-\alpha Ry\} dF_Y(y) = E\left[\exp\{-sY\}\right]_{|s=\alpha R}, \qquad (4)$$

where $F_Y(y)$ denotes the distribution of Y and $E[\cdot]$ denotes the expectation operator. The expectation expressed in the previous equation can be recognized as the Laplace-Stieltjes transform of $F_Y(y)$.

If the interference signal, Y, is assumed to be the sum of I independent random variables, then (4) reduces to

$$E\left[\exp\{-sY\}\right]_{|s=\alpha R} = \prod_{i=1}^I E\left[\exp\{-sY_i\}\right]_{|s=\alpha R} = \prod_{i=1}^I PSR_i = PSR_{Prod}, \qquad (5)$$

where PSR_i obtained by computing the PSR in the presence of the i-th interferer individually as though all other interferers did not exist. We note that the physics of the situation do not seem to support the approach of obtaining total power as the independent sum of powers, although that approach is common in the literature[3].

As a second alternative, if the random variable Y is defined as the maximum of a collection of I interfererers, then, if the target signal is subjected to slow Rayleigh fading, (4) reduces to

$$P\{X/Y > R\} = \int_0^\infty \prod_{i=1}^I P\left\{Y_i < \frac{1}{R}x\right\} dF_X(x). \qquad (6)$$

While certain simplifications of (6) are possible in some very special cases, none of the cases that we have examined are very interesting.

A third alternative is to define the random variable Y as an artificial random variable having average value (power) equal to that of the sum of the I random variables, but having the same distribution as the common distribution of the individual random variables. That is, Y is a scaling of Y_1, the scaling factor being defined by the ratio

$$K_I = (1/E[Y_1])\sum_{i=1}^I E[Y_i].$$

This alternative, which seems to be the most closely aligned with the physics of the situation, reflects what actually happens in the *sum* model for the following reasons. In measuring the effects of interference on packet transmission, we measure the effects of exactly one interferer as a function of SIR. Then, in the presence of multiple interferers, we simply determine the combined power of all interferers, determine the SIR, and then look up PSR on the measured curve for a single interferer.

If the target signal power is drawn from an exponential distribution with parameter α as in slow Rayleigh fading, then we find

$$PSR_{Sum} = P\{(X/Y) > R\} = E\left[\exp\{-sY_1\}\right]_{|s=\alpha K_I R}. \qquad (7)$$

If, in addition, all interference powers are drawn from identical exponential distributions having parameter β, (5) and (7) reduce to

$$PSR_{Prod} = \left[\frac{SIR}{SIR+R}\right]^I \quad \text{and} \quad PSR_{Sum} = \left[\frac{SIR}{SIR+IR}\right], \tag{8}$$

where $SIR = \beta/\alpha$ is the ratio of average signal power, $1/\alpha$, to the average intereference power, $1/\beta$. It is readily established by induction on I that the sum model always provides a larger estimate of PSR than does the product model.

In examining our measurement results, we first considered the sum model, which is represented by (8). Our tests have repeatedly shown that the sum model yields optimistic results. For example Fig. 4 shows how PSR changes with an increase in the number of interfering signals, each of which have $SIR = 10$ dB. The sum model implies that the two interferers with $SIR = 10$ dB each have the same effect as a single interferer with $SIR = 7$ dB. Similarly three, four, and five interferers are approximated by a single interferer having an SIR of 5, 4, and 3 dB, respectively. Fig. 4 also shows the measured PSR for an equivalent single interference; the results are identified in the legend as *sum model*. The optimistic predictions could be explained by the fact that the multiple interference has a combined signal amplitude that is distributed over wider range than a single interferer having the same power. In summary, the results show similar characteristics to those that would be obtained under the threshold model even though the threshold model does not accurately reflect system behavior.

Next, we considered the product model, where PSR in the presence of multiple interferers is calculated accordingly to (5) and PSR_i is the PSR in the presence of i-th interferer only. As pointed out above, this model could result from the case in which the total interference is the result of I independent interferers and the target signal is subjected to slow Rayleigh fading. It also corresponds to a hypothetical situation in which the signal is successively attacked by independent interferers. From the latter perspective, the product model ignores the positive correlation among the effects of different interferers which arises due to the fading process of the desired signal. As the desired signal fades, the probabilities that the target packet is damaged by the other interferers increase simultaneously, and these probabilities also decrease simultaneously as the signal pulls out of the fade. By ignoring this correlation, which appears to be the case in (5), the obtained result becomes too pessimistic. On the other hand, the latter pespective also ignores the fact that the packet can be damaged by the collective effect of the I interferers, which leads to an optimistic drift. The optimistic effect would be dominant for weak interferers, while the pessimistic effect would be dominant for strong interferers. As can be seen in Fig. 4, our measurements show that, in general, the pessimistic effects have more weight overall; values calculated using the product model are lower than the measured values.

It is interesting to note that a special form of the product model was used in [1], and it was identified as a lower bound for the *strongest interference only* model.

Fig. 4 also shows results based on measurements in the presence of two interferers. In the *double-interferer* model, we first sort all interference signals in ascending order $I_1, I_2, ..., I_n$. Then we calculate total PSR according to the formula:

$$PSR = PSR(I_1) \prod_{i=1}^{N-1} \frac{PSR(I_{i+1}, I_i)}{PSR(I_{i+1})}, \tag{9}$$

where $PSR(I_i)$ is the packet-success rate in the presence of interferer i, and $PSR(I_{i+1}, I_i)$ is the packet-success rate in the presence of the two interferers, i and $i+1$, calculated according to (2). The results of this model are shown in Fig. 4 as *double-interferer*. The results obtained by the double-interferer model are pessimistic as in the product model, but they are closer to the measured values than the product model.

5 Simulation Description

The primary objective of our simulation, as was stated earlier, is to calculate the number of successful packets per slot $(SPPS)$ for the network of BRs. The $SPPS$ is a function of the PSR for each transmitting PMU at each BR in a given network, which in turn is a complicated function of the number, velocities and locations of the transmitting PMUs as well as propagation conditions.

In our simulation, we assume a fixed network of BRs whose coordinates, antenna heights and antenna gains are settable parameters. Similarly, the transmit power of PMUs and their antenna gains and heights are settable parameters. Other system parameters that are settable include the standard deviation of the log-normal shadowing distribution, the building penetration loss, the form of the distributions from which the PMU coordinates and speeds of movement are drawn, and in-building and out-of-building mix of the population of PMUs.

The path loss between a PMU and a BR is calculated using a triple-regression model. For distances larger than 1 Km we used the Hata model, which is based on Okumura's measurements[11]. For distances less than 1 km but more than 4 times the BR antenna height, we assumed the free-space propagation model, in which attenuation is proportional to the square of the distance from the BR. For distances less than four times the BR antenna height, we assumed a constant path loss equivalent to the free-space loss at a distance of 4 times the BR antenna height. This constant-path-loss-model is an empirical model related to the typical radiation pattern of the BR antennas. The path loss is further adjusted to account for shadowing and building penetration loss if the PMU is drawn from in-building population.

Once the path loss is determined, we calculate the signal power from each PMU at each BR. Then we calculate the PSR in the absence of interference according to the formula:

$$PSR_n = \begin{cases} 0 & \text{for } P \leq P_{50} - 10 \\ \frac{1}{2}\left\{1 + \sin\left(\frac{\pi}{20}[P - P_{50}]\right)\right\} & \text{for } P_{50} - 10 < P < P_{50} + 10 \\ 1 & \text{for } P_{50} + 10 \leq P \end{cases} \tag{10}$$

where P is the signal power in dBm, and P_{50} is the power at which a PSR_n of 50% is achieved at a given speed. We used $P_{50} = -112$ dBm, -110 dBm, and -108 dBm for speeds 3 Km/h, 32 Km/h, and 100 Km/h respectively. Formula (10), which accounts for fading and noise but not interference, was obtained by fitting a curve to our measurement results.

Similarly, by fitting curves to the measurement results for single interferers we obtained the following formulas for PSR at the various speeds of movement. For 3 Km/h,

$$PSR_i = \begin{cases} \frac{1}{2}\exp\left\{-0.14(5.5 - SIR_i)\right\} & \text{for } SIR_i \leq 5.5 \text{ dB} \\ 1 - \frac{1}{2}\exp\left\{-0.135(SIR_i - 5.5)^{1.25}\right\} & \text{for } SIR_i > 5.5 \text{ dB.} \end{cases} \tag{11}$$

For 32 Km/h,

$$PSR_i = \begin{cases} \frac{1}{2}\exp\left\{-0.175(8.5 - SIR)\right\} & \text{for } SIR \leq 8.5 \text{ dB} \\ 1 - \frac{1}{2}\exp\left\{-0.175(SIR - 8.5)^{1.15}\right\} & \text{for } SIR > 8.5 \text{ dB.} \end{cases} \tag{12}$$

For 100 Km/h,

$$PSR_i = \begin{cases} \frac{1}{2}\exp\left\{-0.18(10.5 - SIR)^{1.2}\right\} & \text{for } SIR \leq 10.5 \text{ dB} \\ 1 - \frac{1}{2}\exp\left\{-0.2(SIR - 10.5)^{1.15}\right\} & \text{for } SIR > 10.5 \text{ dB.} \end{cases} \tag{13}$$

where SIR is equal to the difference between the powers of the target signal and the total interference, both expressed in dBm.

For the multiple interference scenario we tested the sum model, the product model and double-interferer model. For the sum model we added the powers of all interferences (in watts), calculated the SIR in dB from the perspective of a given target transmission, and then used (11) - (13), depending on the speed of movement of the target user, to compute the probability of a successful transmission. For the product model we calculated PSR_i in presence of each interferer and multiplied the resulting PSRs. In the double-interferer model we sorted the interferences by power levels, calculated PSR_i for each of them, then used (2) to calculate the combined effect of two interferences with adjacent power levels, and finally used (9) to determine the combined effects of the multiple interferences.

After calculating the PSR_n in the presence of multiple interferers we multiplied by the PSR_i in presence of noise and fading only in order to obtain the total PSR for the given PMU-BR pair. This is an extension of the product model to the combined effects of noise and multiple interference that has close agreement with our measured data. We use this information to gather statistics on the PSR and the number of hits per packet transmission for the entire network. By repeating this process, we build the matrix $PSR(PMU, BR)$ for all possible PMU-BR pairs.

Although the usual practice is to conduct an experiment at each node that determines whether a particular PMU transmission is successful, our experience has suggested that significant simulation time can be saved by using the following computational formula, which can be shown to yield the same long-term average PSR:

$$PSR(PMU) = 1 - \prod_{BR=1}^{N} [1 - PSR(PMU, BR)], \qquad (14)$$

where N is the total number of BRs. The total expected number of successful transmissions per slot and the expected number of receptions for each transmission, or *hit rate*, are then given by

$$SPPS(P) = \sum_{PMU=1}^{P} PSR(PMU) \quad \text{and} \quad H_j = \frac{1}{P} \sum_{PMU=1}^{P} \sum_{BR=1}^{N} PSR(PMU, BR) \qquad (15)$$

where P is the total number of PMUs transmitting in a given slot.

We repeat the above procedure a large number of times with P fixed. For each sample, the PMU coordinates, speed of movement, shadowing status of the PMU-to-BR path, and the building penetration loss are sampled from their respective distributions. The results of the samples are then averaged to obtain the overall $PSR(PMU)$, $SPPS(P)$ and $H(P)$. Finally, we repeat the entire exercise over the range of P from 1 to a number beyond which increasing P has no effect on $SPPS$.

6 Simulation Results

First, we will consider a network of 16 BRs positioned within 4-by-4 square grid with a separation of 3 Km between adjacent BRs. Each BR has an antenna height of 100 m and an antenna gain of 9 dB. The PMU transmitter power is 1 watt, its antenna height is 1 m, and its antenna gain is -9 dB. One-half of the PMUs are in buildings and are moving at 3 Km/h, 30% are at street level and moving at 32 Km/h, and 20% are on highways moving at 100 Km/h. In-building penetration loss is assumed to be 10 dB, while the shadowing coefficient is drawn from a lognormal distribution having a standard deviation of 8 dB. The distribution of PMU locations is uniform over the 12 Km-square grid.

Fig. 5 shows the change in average $PSR(PMU)$ (averaged over 1000 runs) as function of the number of simultaneous transmission obtained by application of the various models for multiple interference. It can be seen that the sum model produces the most optimistic results, the product model produces the most pessimistic results, and the double-interferer model yields results close to those of the product model. The discrepancy between models is small, especially for a small number of interferers, which is expected to be the typical operational reality.

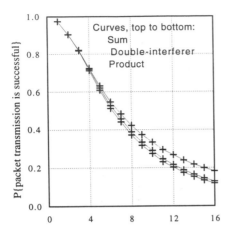

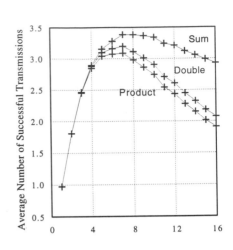

Number of Simultaneous Transmissions

Number of Simultaneous Transmissions

Figure 5: Simulation Results: Probability that packet transmission is successful in the presence of multiple interferers as a function of the number of simultaneous transmissions - comparison of models.

Figure 6: Simulation Results: Packet success rate per slot in the presence of multiple interferers as a function of the number of simultaneous transmissions - comparison of models.

Fig. 6 shows the number of successful transmissions ($SPPS$) for the 16-BR network in packets per slot as a function of the number of simultaneous transmissions. Again, comparison of the results achieved by the various multiple-interferer models shows close agreement, especially for a small number of transmissions. We also find that maximum $SPPS$ is 3-3.5 packets per slot, and this maximum occurs with 7 simultaneous transmissions. This indicates that it is possible to maximize the network throughput by a control mechanism that limits the number of simultaneous transmissions.

Fig. 7 shows how the average number of packet receptions per packet transmission in the given network changes with the number of simultaneous transmissions. For an isolated PMU transmission, more than 3.7 hits can be expected; this can provide valuable information about user location. However, as the number of simultaneous transmissions increases, the number of hits rapidly decreases to below 1 at 5 packets per slot.

We have examined BR-network performance with various input parameters. For example, in Fig. 8 we compare $SPPS$ as a function of the number of transmissions for a 16-BR network in a rectangular grid to that for a 7-BR network with BRs placed in the center and at the vertices of a regular hexagon. The product model and a somewhat different set of parameters is used. Naturally the $SPPS$ is lower for the 7-BR network, and its maximum is reached at a lower number of simultaneous transmissions. Generally we found that the maximum $SPPS$ increases less than linearly with the number of BRs, which is a matter for further study.

In Fig. 8 we show the effects of changing the uniform distribution of PMU locations to a Gaussian distribution having its maximum at the network center. This represents a model of PMU clustering in a downtown area, which is expected to occur in real networks at certain times of the day. Clustering appear to result in a reduction of $SPPS$, and this should be taken into account in network deployment.

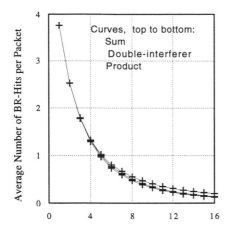

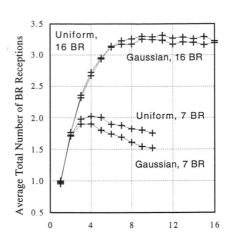

Number of Simultaneous Transmissions

Number of Simultaneous Transmissions

Figure 7: Simulation Results: BR hits per packet in the presence of multiple interferers as a function of the number of simultaneous transmissions comparing 7 and 16 BRs and uniform and Gaussian PMU distributions.

Figure 8: Simulation Results: Packet success rate per slot in the presence of multiple interferers as a function of the number of simultaneous transmissions comparing 7 and 16 BRs and uniform and Gaussian PMU distributions.

We have compared our simulation results with field measurements taken in Dallas in a 5-BR network and found a good agreement with our simulation model with respect to PSR and BR hits. Unfortunately the measurements were done with only a single PMU transmission per slot so that the quality of the results with respect to interference could not be evaluated.

7 Conclusions

This paper describes a simulation model that can be used to evaluate the performance of a system having an arbitrary configuration of the network of BRs and a fixed number of mobile transmitters operating over the same channel and its application to a specific NPCS. The major complexity of the model involves incorporation of the effects of many strong co-channel interferers, which are independently faded during the packet interval. Many other effects contribute to the complexity of the scenario, such as phase noise due to the fading process and the difference in the effects of interference during the synchronization and data transfer phases. We approached the problem of developing a suitable model through a combination of laboratory measurement, field measurement, analysis, curve fitting, and Monte Carlo simulation.

Measurement of the effects of multiple interference, each having an independently drawn power level, is a difficult task. Therefore, we developed models that use either a single or double interference measurement to predict effects of multiple interferences. We investigated models of different complexity levels that produce optimistic or pessimistic results. We found that the models that are popular in the literature are optimistic. Overall we concluded that our models can produce reasonably close estimates of network performance when incorporated into a simulation model.

Perhaps the most important result of the simulation is the maximum $SPPS$ that can be obtained in a given BR network. This type of result can be used to optimize design of a BR network, considering parameters such as BR coordinates and antenna heights, gains, and directivities. The model can also be used to assess the benefits of adding new BRs into a given area, especially if clustering patterns are predicted or experienced in an operational network.

Another use of this model is to predict performance of a collision protocol, such as the slotted Aloha, in the land-mobile radio environment. A more complex protocol would use slotted Aloha for channel reservation only, and a central control station would assign the time slots according to the user location determined on the basis of which BRs receive reservations for a particular PMU. Those topics are a matter for further study.

References

[1] M. Schwartz, *Telecommunications Networks: Protocols, Modeling and Analysis*, Addison-Wesley: New York, 1987.

[2] D. Buchholz, P. Odlyzko, M. Taylor, and R. White, "Wireless In-building Network Architecture and Protocols," *IEEE Network*, Vol. 5, No. 1, November 1991, pp. 31-38.

[3] A. U. H. Sheikh, Y. Yao, X. Wu, "The ALOHA Systems in Shadowed Mobile Radio Channels with Slow or Fast Fading," *IEEE Transactions on Vehicular Technology*, Vol. VT-39, No. 4, November 1990, pp. 289-297.

[4] M. Kocaturk and S. C. Gupta, "Simulation of Co-Channel Interference in Coexisting Cellular TDMA Networks," *IEEE Transactions on Vehicular Technology*, Vol. 43, No 3, Aug. 1994, pp. 753-760.

[5] I. M. I. Habbab, M. Kavenhard, and C. E. W. Sundberg, "ALOHA with Capture Over Slow and Fast Fading Radio Channels with Coding and Diversity," *IEEE Journal on Selected Areas in Communications*, Vol. SAC-7., No. 1, January 1989, pp. 79-88.

[6] S. Ghez, S. Verdu, and S. C. Schwartz. "Stability Properties of Slotted Aloha with Multipacket Reception Capability," *IEEE Transactions on Automatic Control*, Vol. AC-33, July 1988, pp. 640-648.

[7] J. C. Arnbak, W. Van Blitterswijk, "Capacity of Slotted ALOHA in Rayleigh-Fading Channels," *IEEE Journal on Selected Areas in Communications*, Vol. SAC-5, No. 2, February 1987, pp. 261-268.

[8] C. T. Lau and C. Leang, "A Slotted ALOHA Packet Radio System with Multiple Antennas and Receivers," *IEEE Transactions Vehicular Technology*, Vol. VT-39, No. 3, Aug. 1990, pp. 218-226.

[9] D. J. Goodman, Adel A. M. Saleh, "The Near/Far Effect in Local ALOHA Radio Communications," *IEEE Transactions on Vehicular Technology*, VoL. VT-36, No. 1, February 1987, pp. 19-27.

[10] R. Petrovic, W. Rohr, and D. Cameron, "Simulcast in Narrowband PCS", *Proceeding of IEEE MTT-S Symposium on Technologies for Wireless Applications*, February 1995, pp. 65-69.

[11] M. Hata, "Empirical Formula for Propagation Loss in Land Mobile Radio Services," *IEEE Transactions on Vehicular Technology*, Vol. VT-29, No. 3, August 1980, pp. 317-325.

INDEX

A

ACCESS.bus 51
Acolade 196
Adaptive Antennas 89
adaptive equalization 183-194
AMPS 30, 38
amplitude-scaled 164
AM subcarrier 195
antenna 135, 136, 139, 141, 143,
        144, 145
ATIS 195
auto-correlation 171

B

beamforming 89-94
BER 71, 85, 86, 91
  uncoded 169
blocking 65, 66

C

capacity 65, 66
capture models 202, 211
  product model 202, 205, 207,
        209, 210
  strongest interference only model
        207
  strongest interferer only model
        202
  sum model 202, 206, 207, 209, 210
CDMA (code division multiple access)
        38, 59, 60, 71, 81, 87, 92, 104,
        171
CDPD (cellular digital packet data)
        19-23 39, 40, 45, 49
  model 30-33
  forward channel capacity 34-37
  layers 30-33
    MAC 31

MDLP 31
    physical 31
    wireless 32
  reverse channel capacity
        34-37
  transmission failure, definition
        of 30
  simulation 33-38
cell coverage 76
cell plan 153
cell spacing 34
cellular radio 39, 48
Cellu-Trunk™ 2, 4, 5, 6
channel
  dynamics 163
  fading 160
  frequency nonselective 161
chaotic codes 93
co-channel interference 39-50,
        135, 201-205, 211
Coherent Detection 81, 86
computation time 111, 112,
        113, 114, 116
corner effect 147
correlation 136, 140, 142
COST-231 95, 97, 98
coverage 65, 66
cross-correlation 171
CT2 124

D

Delay Lock Loop (DLL)
        81-87
diffraction 99-101
digital sense multiple access
        (DSMA) 39
directional antennas 71, 72,
        74, 75
Direct-Sequence Spread
        Spectrum 81

direct-to-multipath 165
distribtuions
    error-free run 160
    Rayleigh 161
    Rician 162
diversity 89, 90, 123-133
doppler frequency shift 161
downlink 129, 133

E

E&M Signaling 4
error statistics 159
exponential
    curve fitting 168
    functions 164
estimation 163
    Fritchman model 163
    model parameter 166
    SNR 164

F

fading 81, 86
fading channels 183-194
Family A 171
FCC 7, 15, 16, 123
FM 8, 10
forward error correction 201
forward link 86
frequency hopping 129

G

GMSK 196
Gold 171
GPS 96, 128
GSM 104, 123, 129
GTD 95, 98-101

H

handoff 147
handoff criterion 150
HDTV 7

Hilbert Space 164
Hopfield network 93
horizontal polarization
        123, 125, 127

I

illumination area 111, 113,
        114
imaging method 111, 112,
        113, 116
independent service provider
        29
Intelligent Transportation System.
        See ITS
interference 15, 102-106
interference limited 135, 136,
        140, 142, 145
interference models 211
    double-interferer model 207,
        208, 210
    multiple interferer model 210
IPL 102, 105
IS-95 59, 62
ISDN 7
ITS, ITS Architecture 27
    communication system
        definition 28
    data dictionary 28
    independent service provider
        subsystem 29, 31
    ISP-Vehicle communications
        30, 33, 35-37
    logical architecture 28
    physical architecture 28
    vehicle subsystem 29, 31

L

least-squares fit 145
link budget 9, 11
LMDS 7-17
LMS Algorithm 89, 92
Loral Federal Systems 27

M

Markov modeling techniques
        163
matrix transition probability
        160
maximal ratio combining 62, 64,
        65, 126
maximum likelihood sequence
        estimation 183-194
microcells 147, 152
microwave 95, 106
millimeter wave 7
mobile data base station
        (MDBS) 39
Mobile Telephone Switching
        Office (MTSO) 1, 3, 4, 5
model 159
    Fritchman 159
    Hidden-Markov 159
    Markov 159
    open areas 162
    Simple-Partitioned Fritchman 160
    urban areas 161
multimedia 7
multipath fading 39, 40

N

narrowband 201-203, 211
NCCG 82, 83, 86
network capacity 74
noise additive white gaussian 161
noise limited 136, 139, 142

O

OFS 95, 96, 101-106
OPNET 27, 31 32, 37
outage 65, 66
outage probability 135, 136,
        138, 139, 140,142,
        143, 144, 145
outer cell interference 74

P

PCS 7, 95-106
PCS-1900 123-133
periodic distribution 32
personal area network 51
phase
    2-phase 171
    4-phase 171
phase-shift 161
pilot symbol 81, 86
PN Code 81
Poisson distribution 32
polarization 8, 13
polarization diversity 123-126,
        133
power control 129
Private Branch Exchange
        (PBX) 1, 2, 3, 4
projection
    orthogonal 165
    theorem 164
propagation measurement
        12-15, 96
PSK modulation 102

Q

quasi-static 60

R

radio propagation 111, 115,
        116
RAKE 59, 60, 65
Rayleigh fading 202-207
ray tracing 111, 113, 116
results 166
reuse factor 136, 139
reverse link 71, 72, 86
Routing Correlator 3, 4, 5

S

satellites 15

scattering 101
serial bus 51
shadowing 136
signal
  diffused path 162
  finite-energy 164
  line-of-sight 162
  measured 164
  quality indicator 147, 150
  reference 164
simulations 85, 91
  Monte-Carlo 159
simulcast 201, 203
slotted Aloha 201, 203, 212
  with reservation 201
soft handoff 59, 62, 63, 64,
        66, 67, 147, 150
space diversity 123, 124,
        127, 133
SQI 147, 150

T

TCP/IP 34, 38
TDMA 38, 104, 129
threshold 138, 140, 202
TIA 106
tilt angle 135, 136, 138, 139,
        140, 141, 143
time delay 161, 164
tracking jitter 81-87
transmission latency 34

U

UDP/IP 38
uplink 128-133
Urbansville scenario 27, 33
US DOT 27
user services 27

V

vector column 160
vector row 160

vehicle 29
vertical polarization 123, 126, 127

W

Wired-to-Wireless Adapter
        (WWA) 3, 4
Wireless data networks 19-21
Wireless local area network
        19-21, 23-26
Wireless Local Loop 71, 72